Implementing the ISO/IEC 27001 Information Security Management System Standard

Third Edition

For a complete listing of titles in the
Artech House Information Security and Privacy Series
turn to the back of this book.

Implementing the ISO/IEC 27001 Information Security Management System Standard

Third Edition

Edward Humphreys

ARTECH HOUSE
BOSTON | LONDON
artechhouse.com

Library of Congress Cataloging-in-Publication Data
A catalog record for this book is available from the U.S. Library of Congress.

British Library Cataloguing in Publication Data
A catalogue record for this book is available from the British Library.

Cover design by Andy Meaden

ISBN 13: 978-1-68569-075-5

685 Canton Street
Norwood, MA 02062

10 9 8 7 6 5 4 3 2 1

To my father and mother (Thomas Humphreys and Alice Humphreys), my sons (Alexander, Thomas and James), my grandchildren (Siena, Juliet, Penelope, Georgia and Charles) and, finally, to my wife Angela Elizabeth

Contents

Acknowledgements *xiii*

Introduction *xv*

1 Information Security **1**

1.1 The Importance of Being Informed 1
1.2 Globally Connected 3
1.3 More Ado About Risks 4
1.4 Decoding the Secret of Information Security Management 5
1.5 Management and Awareness 6
1.6 Legislation, Regulation and Governance 7
1.7 En Route to a Certified Business Environment 9

2 ISO/IEC 27000 Series **11**

2.1 ISO/IEC Standardization 11
2.1.1 Overview 11
2.1.2 ISO/IEC JTC 1/SC 27 12
2.2 Overview of the ISO/IEC 27001 Standard 12
2.2.1 International Standards 12
2.2.2 ISO/IEC 27001 and Related Standards Series 13
2.2.3 Standards Related to the ISMS Series 15
2.3 Short History of ISO/IEC 27001 17
2.3.1 Emergence of ISMS 17
2.3.2 Internationalization 19
2.4 Overview of ISO/IEC 27001: 2022 19
2.4.1 Introduction 19
2.4.2 ISMS Audience 20
2.4.3 Mandatory Statements of Requirement 20
2.4.4 Processes 21
2.4.5 ISMS Stages 21
2.4.6 Organizational and Top Management Focus 22
2.4.7 Support 22

2.4.8 Risk-Based Approach 22
2.4.9 Operations 23
2.4.10 Performance Evaluation and Continual Improvement 23
2.4.11 Annex A 23
2.5 Conformance to ISO/IEC 27001 23

3 ISMS Business Case and Context **25**
3.1 ISMS Business Case 25
3.2 Organizational Context 26
3.2.1 Understanding the Business 26
3.2.2 Internal Issues and Context 27
3.2.3 External Issues and Context 28
3.3 Needs and Expectations 30
3.3.1 Interested Parties 30
3.3.2 Requirements Relevant to the ISMS 30
3.3.3 Gathering Requirements Relevant to the ISMS 34
3.4 ISMS Scope 35
3.4.1 What to Consider and What to Include 35
3.4.2 Object of ISMS Scope 36
3.4.3 Defining the ISMS Scope 37
3.4.4 Scope Example 38
3.4.5 External and Internal Connections 40

4 Managing the ISMS Risks **41**
4.1 Importance of Risk and Opportunity 41
4.1.1 Risk Definition 41
4.1.2 Opportunity 42
4.1.3 Risk Attitude, Tolerance and Appetite 42
4.1.4 Information Security Risk Appetite and Tolerance 43
4.1.5 ISMS Risks 44
4.2 Risk Management Process 45
4.2.1 Changes in the Process 45
4.2.2 Risk Assessment 45
4.2.3 Risk Treatment 52
4.2.4 Determine the Controls 57
4.2.5 Statement of Applicability 57
4.2.6 Risk Treatment Plan 58
4.2.7 Risk Owner's Duties 58
4.3 Ongoing Reassessment of Risk 58
4.3.1 Risk Reviews and Reassessments 58
4.3.2 Risk Monitoring 59
4.3.3 Updating the Risk Treatment 59

5 ISMS Leadership and Support **61**
5.1 Management Policy 61
5.1.1 Approval, Communication and Awareness 62
5.1.2 Policy Review 63
5.1.3 Management Policy Sets the Scene 63

5.2 Leadership 63
5.3 Roles and Responsibilities 64
5.4 Resources 65
5.5 Training and Awareness 67
5.5.1 When Should Training Take Place? 67
5.5.2 Training Methods 67
5.5.3 ISMS-Related Topics 68

6 Measures to Modify the Risks **71**
6.1 Determining the Controls 71
6.1.1 Control Framework 71
6.1.2 Process of Determining a Control Set 72
6.1.3 Control Sources 74
6.1.4 Existing Control Sets 75
6.2 System of Controls 76
6.2.1 Control Framework 76
6.2.2 Requirements and System of Controls 76
6.2.3 Examples of Groups of Control 77
6.3 Control Through Policies and Procedures 80
6.3.1 General 80
6.3.2 Policies and Procedures 81
6.3.3 Approval, Communications and Awareness 82
6.3.4 Review 83
6.4 Examples of Control Sketches 83
6.4.1 Overview 83
6.4.2 Information-Handling Procedures 83
6.4.3 Access Control Policy, Procedures and Process 92
6.4.4 Acceptable Use of Resources 96
6.4.5 Human Resource Policy, Procedures and Process 99
6.5 Sector and Application Specific Controls 101
6.6 ISO/IEC 27001 Annex A 103
6.6.1 Purpose and Definition 103

7 ISMS Operations **107**
7.1 Operational Planning, Management and Control 107
7.1.1 Processes 107
7.1.2 Example Processes 108
7.1.3 Training, Awareness and Usage 109
7.2 Ongoing Risk Assessment 109
7.3 Ongoing Risk Treatment 110
7.4 Example Sketches of Operational Threats 111
7.4.1 Malware 111
7.4.2 Unauthorized Access 113
7.4.3 Insider Threat 114
7.4.4 System Availability 116
7.4.5 Social Engineering 118
7.5 Examples of Sketches of Operational Processes 119
7.5.1 Protecting Information in the Operational Environment 119

7.5.2 Backups 120
7.5.3 Capacity Planning 123
7.5.4 Change Management 124
7.5.5 Third-Party Logistics and Supply Chain Management 125
7.5.6 Third-Party Services 125
7.6 Incident Management Narrative 127
7.6.1 Events That Compromise 127
7.6.2 Use Cases 128
7.6.3 Processes 129
7.6.4 Incident Investigations 134
7.6.5 Incident: Cause and Effects 134
7.6.6 Incident Management Team 135
7.6.7 Standards 136
7.7 ISMS Availability and Business Continuity Narrative 136
7.7.1 Value and Importance 136
7.7.2 Business Impact 137
7.7.3 Plans 138
7.7.4 Processes 139
7.7.5 Standards 139
7.8 ISMS Use Examples 140
7.8.1 SME Design Services 140
7.8.2 Legal Services 141
7.8.3 Electronic Accounting System 142
7.8.4 Government Payment System 143
7.8.5 Outsourcing Call Centre Operations 144
7.8.6 Manufacturing Systems 146
7.8.7 Supply Chain Management 147

8 Performance Evaluation 149
8.1 Performance, Change and Improvement 149
8.1.1 How Effective, Adequate and Suitable Is the ISMS? 149
8.1.2 Change and the Certainty of Change 150
8.1.3 Change Management 151
8.1.4 Tracking and Reviewing Ongoing Change 153
8.1.5 Informed Decision-Making 154
8.2 Monitoring and Operational Reviews 155
8.2.1 Monitoring 155
8.2.2 Monitoring and Review of Staff Awareness, Competency and Use of the ISMS 157
8.2.3 Monitoring and Review of Information Security Processes 159
8.2.4 Monitoring and Review of Information Security Controls 161
8.2.5 Monitoring and Review of Technology and Network Services and Infrastructure 162
8.2.6 Monitoring and Reviewing Third-Party Contracts and Services 163
8.2.7 Monitoring and Review of Legal and Contractual Compliance 164
8.3 ISMS Measurements Programme 165
8.3.1 ISMS Metrics and Measurements 165
8.3.2 Measurement Programme 168
8.4 Ongoing Risk Management 177

8.4.1 Risk Responsiveness and Commitment 177
8.4.2 Regular (Planned) Risk Assessments 178
8.4.3 Risk Measurements and Metrics 179
8.4.4 Risk Management Reviews 180
8.4.5 Information Security Incident Management 181
8.5 ISMS Internal Audits 181
8.6 Management Reviews of the ISMS 182
8.6.1 Management Review 182
8.6.2 Input for the Management Review 183
8.6.3 Output of the Management Review 184
8.7 Awareness and Communications 185

9 Improvements to the ISMS **187**
9.1 Continual Improvement 187
9.1.1 Improvement 187
9.1.2 Maintaining Effectiveness, Suitability and Adequacy 189
9.1.3 Holistic Effectiveness 189
9.2 Conformance and Nonconformance 190
9.2.1 Nonconformity 190
9.2.2 Corrections 190
9.2.3 Corrective Actions and Root Causes 190
9.2.4 Some Common Causes on Nonconformity 191
9.2.5 Case Study 1 192
9.2.6 Case Study 2 192
9.2.7 Case Study 3 193
9.3 Making Improvements 193
9.3.1 Planning and Implementing Improvements 193
9.3.2 Implementing Improvements to Management Leadership and Commitment 194
9.3.3 Implementing Improvements to Resources and Competence 194
9.3.4 Implementing Improvements to Awareness and Training 194
9.3.5 Improvements to Processes 194
9.3.6 Improvements to Policies and Procedures 195

10 Accredited ISMS Certification **197**
10.1 Overview 197
10.2 International Certification 198
10.2.1 Global Take-Up 198
10.2.2 Motivation 199
10.2.3 Costs and Resources 199
10.3 Certification and Accreditation 201
10.3.1 ISO Standards and Definitions 201
10.3.2 Interested Parties Involved in Certification 202
10.3.3 Accreditation 203
10.3.4 Certification 203
10.3.5 Selecting an Accredited CB 204
10.4 Standards Involved 204
10.4.1 Accreditation 204
10.4.2 Certification 205

10.4.3 Supporting Auditing Guidance (Not Mandatory) 205
10.4.4 End-User Organizations (ISMS Owners) 205
10.5 ISMS Audits 205
10.5.1 Certification Scope 205
10.5.2 Audit Process 206
10.5.3 Nonconformities 208
10.5.4 Audit Report 208
10.5.5 Surveillance Audits 208
10.5.6 Recertification 209
10.5.7 Audit Trails 210
10.5.8 Competence 211

11 Coda **213**
11.1 The ISMS: A Living System 213
11.2 Processes Cycles and Sequences of Activity 214
11.2.1 Information Life Cycle 214
11.2.2 Risk Management 215
11.2.3 ISMS Performance and Improvements 215
11.3 ISMS the Business Enabler 216

Bibliography **219**

About the Author **225**

Index **227**

Acknowledgements

I would like to thank the reviewer for their suggestions and invaluable comments. I would also like to thank the SC 27 colleagues for their supportive suggestions.

Introduction

Information security is a global issue affecting systems, processes and services used for international trading, electronic commerce, communications, social media and national infrastructures. Managing information security is an even more critical issue because it involves using and managing the policies, procedures, processes, control measures and supporting services and technologies that are needed to protect the information that organizations, governments, consumers and citizens rely on to carry out their business and to live their lives. Information security management needs to be effective, suitable and appropriate in order to protect information from the risks that business and society faces in this digital age. For example, processing information is essential for those systems and services on which we rely to provide transportation, telecommunications, healthcare, the supply of energy (such as electricity, water and gas) and many other common things we accept as being there when we want them. The information systems that provide vital management support for these different aspects of this public infrastructure are vulnerable to a wide range of information security risks.

Our dependency on information in digital form is all pervasive and ubiquitous, as are the risks to this information. Information could be disclosed and accessible to unauthorized users, corrupted or modified either in some unauthorized or accidental way or lost or unavailable due to a system failure or disruption. These represent the three important security objectives: protecting the confidentiality, integrity and availability of information. Understanding what the risks are and assessing how these risks affect and impact business in terms of these objectives are central to being able to manage these risks effectively. An organization needs to assess its risks in terms of the potential impact of a security incident on its business and the likelihood of this security incident occurring. It needs to adopt an approach to risk assessment that is effective, suitable and appropriate to its business.

The simple message to all organizations and governments is that investing in the management of information security risk is a wise investment.

All organizations are dependent more than ever on information and technology for driving their business processes, productivity and efficiency and so the need to protect this information is vital to their survival in today's digital driven economy. Assurance that the confidentiality, integrity and availability of this information as well as protecting personal data is at the core of business success.

This book covers the requirements of the international standard ISO/IEC 27001 on information security management. This standard was developed around 30 years ago to meet the needs of the international business market to protect one of the important assets: information, a valuable currency and commodity for being able to survive and prosper in today's dynamic business world.

This book particularly looks at edition three of ISO/IEC 27001 (2022). Several narratives giving explanation and practical advice are provided to illustrate the use of these standards in business situations. It includes what is involved in establishing and designing, implementing and deploying, monitoring and reviewing and updating and improving an information security management system (ISMS). Using a well-tried and tested process model that is embedded in ISO/IEC 27001, the key ISMS standard, not only enables a business to establish an effective information security regime and risk management process and to deploy a number of management controls to continually improve the ISMS to ensure that they still have effective information security. The process of third-party certification is covered, which enables businesses to demonstrate through independent audits that they have in place an ISMS that effectively protects their information.

The author cannot be held legally responsible or in any way liable for the real-world use, application, implementation or interpretation of the information given in this book by users, companies, businesses, organizations or any other third parties. Whilst every care has been taken in developing and producing the material in the book, the author accepts no liability for any loss, damage or harm caused, arising directly or indirectly in connection with the reliance on this book material. As the standards referenced in this book and their content will change over time, the material in this book will over time become out of date; therefore, users need to refer to the latest versions of the standards. Therefore, users and third parties of the material in the book are wholly liable and responsible for the correct use, application, implementation or interpretation of the information given in this book.

Finally, the spelling of words used in this book are the internationally accepted spellings adopted and used in ISO and ISO/IEC standards and are from the *Oxford English Dictionary* (OED).

CHAPTER

1

Contents

1.1 The Importance of Being Informed

1.2 Globally Connected

1.3 More Ado About Risks

1.4 Decoding the Secret of Information Security Management

1.5 Management and Awareness

1.6 Legislation, Regulation and Governance

1.7 En Route to a Certified Business Environment

Information Security

1.1 The Importance of Being Informed

Do nothing secretly; for time sees and hears all things, and discloses all.

—Accredited to Socrates

Historically information[1] has always played an important role from ancient times to the present day, for example:

- Its use in the military campaigns of Alexander III of Macedon, Xerxes, Julius Caesar and others, to its use in managing the societies and cultures of ancient Greece, Egypt, Persia and China, and across ancient cultures, and its use to help the Phoenicians, Greeks, Persians and Romans trade across their empires;
- Shaping our views and thoughts of the world through progressing our knowledge and understanding of mathematics, science, medicine and philosophy from ancient times to the modern era;

1. Informare (to give form to the mind). πληροφορα from πλρης (fully conveys).

- The silk and spice routes, involving countries in Europe and the Near East, Middle East and Far East, trade and navigation of the Renaissance merchants and seafarers, all were dependent on the availability and accuracy of information;
- The invention of the printing press brought information to many, and the telegraph system got information to others faster across vast distances, and today's internet and broadband technologies information gets to even more people, even faster;
- Twenty-first century with its development cyber warfare, Big Data technology, Internet of Things (IoT), artificial intelligence (AI) and machine learning, cloud services, Smart Cities, social media and personally identifiable information (PII).

The list of examples on the use and importance of information is as endless as history itself. Today we say we live in a globally connected world that is constantly expanding and shaping the way we think, work, act and play.

Information from ancient times to now (and in the future) has and will continue to be a valuable asset to have and possess and so it has always been and will continue to be the target of risk:

- Risks to information (e.g., disclosure and theft of information about their personal estate and affairs, their health, their discoveries and inventions, modification of their plans, loss or corruption of their personal data, protecting image and reputation);
- Risks from information used by their opponents (e.g., a competitor knows more than they do, traders in the Renaissance times knew a better trading route without getting lost or falling foul of rocks and uncharted waters,[2] Sun Tzu's Art of War and business strategy);
- Risks from sharing information with others (e.g., a partnership, business contracts, relationships, social media).

The very same market innovations, products and services that have shaped this digital world of which we are a part and the benefits we have

2. The Spanish and Portuguese explorers and seafarers used the notion of risk in the sixteenth and seventeenth centuries to refer to sailing into uncharted waters. This use of risk is associated with a notion of uncertainty related to space. We then adopted a notion of risk related to time, as used in financial processing and investment. In this latter case, it is related to the probable consequences of investment decisions for borrowers and lenders. Today the notion of risk relates to a wide range of other space/time situations of uncertainty, including risks to our organization's financial and nonfinancial assets.

gained from this digital world, have also given us a world where the information we use, value and rely on is at risk and is becoming more difficult to protect. The importance and value of information and protecting this value are the central themes of this book.

1.2 Globally Connected

Thought is an idea in transit, which, when once released, never can be lured back, nor the spoken word recalled. Nor ever the overt act be erased.

—Accredited to Pythagoras

Today's Information Society, digital economy, cyber world or whatever other name we might choose to describe the phenomena that are dominating our world is increasingly driving, shaping and reforming the way we work, live and play. Organizations around the world are doing business across public and private networks. Every type of business, from the very small to the very big, is affected by the Information Society. Citizens in many countries of the world are connecting to the internet doing online shopping and banking and hotel and airline reservations. Industry and governments alike are seeing the benefits of going down an electronic commerce route and reaping the benefits of the digital economy.

Today's technologies make it easier to have access to information, to store more of it, to process it faster and to communicate it globally in seconds. Mobile and smart technologies make it easier to stay globally connected all the time and the exponential growth in mobile apps is transforming our lives like never before, enabling us to do more on the move. This capability gives us the means of being more informed and more capable of exploiting and taking advantage of a vast range of business and private opportunities. However, at the same time, the same technology can be used to damage or destroy information.

With the emergence of the IoT, it is not just users via laptops and mobile and smart devices that are connected but many billions of small devices and sensors and other things connected through wireless networks have also connected to the internet making the size of the internet community bigger. These things are helping, for example, to control smart energy, transport and building systems and so are collectively involved in sensing, communicating and, in a limited way, processing information, which may be critical, sensitive or even personal.

The widespread use of technologies such as AI and machine learning provides many opportunities in our daily lives, in business and society as a whole, and this is reshaping human environments beyond whatever we have known in the past.

The diverse range of network technologies and services available today make it easy to access the world, to be globally connected and to be online at any time, from almost anywhere. This makes information more vulnerable whether it is business, government, personnel or private information.

The email to a friend or business colleague, once sent, is public; it can be deleted by the receiver, but it cannot be recalled. Remarks on a social media site can become viral within a short period of time. The power of the internet and global interconnectivity has given many benefits to our work and our lifestyles; it has also made many of things we do more public and less private, it has raised the level of risk our world is facing and it has raised the need for more awareness, care and attention to how we should use this global connectivity and how we should behave in a responsible, safe and secure way.

1.3 More Ado About Risks

As a rule, men worry more about what they can't see than about what they can.

—Accredited to Julius Caesar

Risk management is an old concept practiced by the Ašipu monks in the Tigris and Euphrates delta around 3200 BCE. The risk-related problems they addressed were related to finance, healthcare and simple technologies. The idea of risk management has been used throughout history since the time of the Ašipu monks to control and mitigate against the effects of risk.

Today we address risk as the effect of uncertainty on objectives as defined in the standard ISO 31000; this is not dissimilar to that of the Ašipu monks. What is the uncertainty of whether your organization is going to be infected by a malware attack? If it happens, what is the effect on the organization's security objectives? Maybe information will be stolen by competitors which will have a financial impact on the organization or maybe information is modified or destroyed, and the impact is loss of productivity whilst the information is restored.

Organizations' dependency, use and application of information are all pervasive; all information technology (IT)-based business processes are driven by the supply of information and the output of this process is yet again information. This dependency on information is made more urgent by the potential risks to this information and the uncertainty of the risk. Understanding what the information security risks are and assessing how these risks have an effect and impact business are vital to being able to manage these risks effectively.

An organization needs to assess its risks in terms of the impacts resulting from a security incident happening. It needs to adopt an approach to risk assessment that is suitable to its business environment and to invest in the management of risk and to minimize any business impact. The risk management process needs to minimize the risks (effects of downside risks) by making well-informed decisions of how to treat the risks that is most cost-effective whilst taking account of the need to support and maximize the business opportunities (opportunities through upside risks). The risk management process involves taking the necessary action to ensure that the options to treat the risks are implemented whilst supporting the business opportunities.

It is important that an organization deals with information security at all levels to ensure availability and continuity of business resources, to treat the risks to avoid or reduce any potential damage and impact to the business.

1.4 Decoding the Secret of Information Security Management

The secret of change is to focus all your energy, not on fighting the old but on building the new.

—Accredited to Socrates

Management systems such as information security management system (ISMS) are based on the cyclic model of continual improvement in order to deal with changes in our business environment and to maintain a level of effectiveness suitable and appropriate to the organization and its business. The ISMS ethos is about risk management, monitoring and reviewing the risks and the effectiveness of their treatment, keeping your risk treatment up-to-date, and to make continual improvements to the treat the risks associated with changes.

Uncertainty and change are an inevitable, inescapable, universal aspect of life and the world in which we live, whether in business or personal life there are uncertainties and there are changes: "nothing is as permanent as change."[3] The definition of risk as discussed in Section 1.3 revolves around the notion of uncertainty and all changes (expected or unexpected), which all carry an element of risk. Information security risks threaten the digital world in which we live and, as this digital world evolves and changes, the risks change and evolve. As businesses and consumers engage in this digital world, they need to confront the uncertainties of the risks that are involved. The "secret" of information security management systems lies in the continual improvement processes involved, which enable businesses to manage the changes and risks and keep up-to-date with their risk threat

3. Accredited to Heraclitus, "All is flux, nothing stays fixed."

programme, applying effective security processes and controls to protect their information. This means preserving the confidentiality, integrity and availability of information by managing the risks. These three security objectives should be the focal point of the organization's information security policy. In addition, the organization needs to take care of the personal data or PII; it processes relating to its staff, customers and consumers. The ISMS continually evolve and improve with the pace of change to achieve these objectives.

The digital revolution as a way of doing business has been growing by leaps and bounds and it has become the standard way of doing business. Technologies such as cloud, Big Data, IoT and AI are making our world more connected, smarter and more risky. The standard ISO/IEC 27001 provides the means of managing the information security risks in our cyber world. This standard, if properly implemented, can give support to addressing questions such as: "Is my organization fit to do online business in a secure way?" "Is my organization's online site secure?" "Do I have trust and confidence in my trading partners' ability to keep information safe and secure?"

There are questions related to confidence in third-party arrangements:

- "Are they fit to have access to my organization's information processing facilities and to share information with?"
- "Are they fit to look after my organization's information and information processing facilities?"

1.5 Management and Awareness

Education is kindling the flame, not filling the vessel.

—Accredited to Socrates

The emphasis of ISO/IEC 27001 is the protection of information and what underpins the success of having an effective, suitable and appropriate ISMS implementation is management. This entails the scope of management in its broadest sense; from the CEO and others in the C-suite category through other layers of management, it involves the management of people, policy and procedures, processes, systems of risk management controls and information and to be a comprehensive framework for managing information security it also includes the management of everything that supports the business: the services, business applications and IT infrastructure.

Management is the heart and soul of a successful ISMS implementation. Direction needs to be driven by top management policy, there needs to be active commitment and leadership for the ISMS and this needs to come from the top and resources need to be made available and committed to from the top of the organization. Management needs to engender and promote an information security culture in the organization to support its use of its ISMS.

Awareness and education are also critical to having an ISMS that will be effective in managing the risks. Using the organization's ISMS in practice, during the day, every day, every week of the month, should be an integral part of the managers and employees' work function. Awareness and education should, in this sense, be about the practice of using the ISMS not about the theory of the ISMS. For managers and employees, it should be about using and applying the ISMS policies and procedures to their particular work circumstances. What should they do and how should they respond to security events as they happen? What is the proper way to secure information they handle? Do they know which procedure to use for a particular work task that involves some security aspect? These and many other day-to-day practical aspects of carrying out their specific job function in a manner that will protect the information with which they are dealing is at the heart of good information security awareness. So it is not just about building up one's knowledge and understanding but how to put this knowledge in practice, knowing how to do things in practice and knowing how to react and respond to information security risks. This practical understanding might make the difference between the organization successfully surviving a malicious cyber-attack or suffering untold damage because of this attack.

Information security is teamwork, and everybody plays an important role in managing the risks; everybody from top management down to employees should all be aware of their responsibilities and what they need to do and to contribute towards achieving an effective ISMS.

1.6 Legislation, Regulation and Governance

Good people don't need laws to tell them how to act responsibly ... and bad people will find a way around the laws.

—Accredited to Plato

Over the past three decades or so, more attention has been given to the development of laws and regulations related to some aspect of information security and privacy, for example, data protection and privacy legislation, computer misuse and hacking, social media, electronic commerce and the use of electronic signatures, digital evidence or cyber security. It is very likely that the future will see more such regulations appear with more technological developments on the horizon.

Laws and regulations have a business impact on any organization as they clearly need to comply with such legislation and also to demonstrate their compliance to such legislation.

The growth in cyber-crime has further increased the need for legal protection to protect businesses and their transactions as well as personal data that is stored and processed by servers in different jurisdictions.

Some examples of current legislation include:

- The UK Computer Misuse Act (1990) deals with criminal activities to criminalize unauthorized access to computer systems; DPA (Data Protection Act 2018), UK-GDPR (UK General Data Protection Regulation), NIS Regulations (Network and Information Security Regulations 2018), Telecommunications (Security) Act 2021, UK eIDAS (Electronic Identification and Trust Services for Electronic Transactions Regulations 2016) and PECR (Privacy and Electronic Communications Regulations).
- The EU Cybersecurity Act (2019), EU Cyber Resilience Act, EU Cyber Solidarity Act and EU Directive on Data Protection (plus the legislation of each Member State, their national interpretation of the EU Directive), which covers the gathering, handling and exchange of personal data.
- Data Protection/Data Privacy legislation of non-European Union (EU) countries include Australia, Canada, Hong Kong, Malaysia, Mexico, New Zealand, Norway, Republic of South Korea, Singapore, South Africa and Switzerland.
- The EU Directive on Electronic Signatures provides legal provisions regarding the legal admissibility of electronic signatures for example, those used for online transactions, and other similar pieces of legislation, appear in other parts of the world such as in Hong Kong and Japan.
- Cyber security-related regulations in the United States include the 1986 Computer Fraud and Abuse Act, the 1996 Health Insurance Portability and Accountability Act (HIPAA), Children's Online Privacy Protection Act (COPPA) of 1998, the 1999 Gramm-Leach-Bliley Act (GLBA), The Federal Information Security Management Act (FISMA) of 2002, Homeland Security Act and the 2013 Cyber Intelligence Sharing and Protection Act (CISPA) and The Cybersecurity Information Sharing Act (CISA) of 2015.

- SoX or Sarbox (Sarbanes Oxley) also known as the Public Company Accounting Reform and Investor Protection Act of 2002.
- Cybercrime laws: over 80% of countries have enacted legislation to combat cybercrime.
- International Copyright and Licensing Laws, which provide legal measures against piracy and theft of intellectual property.

1.7 En Route to a Certified Business Environment

We acquire a particular quality by acting in a particular way.
—Accredited to Aristotle

The quality of our information security depends on the actions management takes to achieve an effective ISMS. It is estimated that organizations in over 120 countries around the world organizations are using ISO/IEC 27001 to manage their information security by implementing an ISMS. In addition, many of these organizations have opted for having their ISO/IEC 27001 ISMS certified by a third-party certification body. This certification process against ISO/IEC 27001 provides a common international benchmark for assessing an organization's information security management system. Certification assures and gives confidence to management, customers and suppliers and business partners that an organization is applying best information security management practice. It has been shown that certification opens new avenues of business opportunity for those organizations that can demonstrate that they have a protected business environment.

ISO/IEC 27001 and the process of certification provides a tried and tested framework and a metric, evaluation and monitoring system for determining whether your business is "fit for purpose" from an information security management perspective. The certification audit will check that the ISMS is in compliance with the requirements in ISO/IEC 27001.

Establishing a certified ISO/IEC 27001 business environment to demonstrate confidence to the management board, top management, C-suite managers and your business partners and customers that your organization is "fit to do business with" is common among many thousands of organizations around the world: small, medium and large-sized organizations in most business and industry sectors as well as governments and academic and research institutions.

CHAPTER

2

Contents

2.1 ISO/IEC Standardization

2.2 Overview of the ISO/IEC 27001 Standard

2.3 Short History of ISO/IEC 27001

2.4 Overview of ISO/IEC 27001: 2022

2.5 Conformance to ISO/IEC 27001

ISO/IEC 27000 Series

Choose always the way that seems the best, however rough it may be; custom will soon render it easy and agreeable.

—Accredited to Pythagoras

2.1 ISO/IEC Standardization

2.1.1 Overview

The work in the International Organization for Standardization (ISO) and the International Electrical Committee (IEC) is carried out by Technical Committees (TCs) and Subcommittees (SCs). These committees are responsible for the executive decision-making and overall management of the standards programme. Working groups are established in the SCs to carry out the development of the standards. In addition, the Joint Technical Committee (JTC 1) is a joint ISO and IEC committee responsible for IT-related standards. For more detailed information about the standardization process, who gets involved in the work, the technical committees and what they do and the deliverables (standards, technical specifications and reports), the reader is directed to the ISO website, https://www.iso.org/iso/developing-standards.html.

This chapter provides an overview of the work in ISO/IEC JTC 1 related to information and IT security standardization, in particular, ISMS standards and the subcommittee SC 27.

2.1.2 ISO/IEC JTC 1/SC 27

2.1.2.1 SC 27 Structure

ISO/IEC JTC 1/SC 27 is the committee responsible for a wide range of information security, cybersecurity and privacy protection standards projects. The subcommittee consists of five working groups (WG 1 to WG 5) covering a diverse range of projects:

- WG 1 Information security management systems;
- WG 2 Cryptography and security mechanisms;
- WG 3 Security evaluation, testing and specification;
- WG 4 Security controls and services;
- WG 5 Identity management and privacy technologies.

2.1.2.2 SC 27 Information

More information about SC 27 can be found at https://www.iso.org/committee/45306.html: this includes membership, co-operating bodies, projects and published standards. Also, SC 27 has its own website https://committee.iso.org/home/jtc1sc27.

SC 27 has a large number of members and co-operating bodies (liaisons), which means it receives a wealth of contributions and knowledge from professionals in the field of information security from around the globe. This collective engagement with the international community means the needs and expectations of organizations and businesses working in most market sectors contribute to the development of the final product: the standards. This wealth of input, as well as the review and feedback of contributions, adds to the value of the standards developed as it reflects a true global view of requirements and solutions for information security, cyber security and privacy protection.

2.2 Overview of the ISO/IEC 27001 Standard

2.2.1 International Standards

The ISO/IEC 27001 set of ISMS standards is developed and maintained in SC 27/WG 1. The set of ISMS standards is produced at the international level in response to market needs, to protect global trading, to mitigate against cyber risks and to meet various GRC (Governance, Risk and Compliance) requirements. Like all ISO and IEC standards, the ISO/IEC 27001 standards are based on global expert opinion and developed through a multistakeholder process, using a consensus-based approach. All such standards go through a maintenance life cycle to keep them current and up-to-date; this means that standards are often revised after a period, normally around 5 years. Therefore, a standard may have many editions, each new edition replaces the previous edition. Users of ISO/IEC standards should always check if the version they are using is the most current edition.

This book addresses the requirements of the third edition of ISO/IEC 27001 and how they can be implemented. The following is the history of ISO/IEC 27001:

- 2005: The first edition of ISO/IEC 27001 was published.
- 2013: The second edition of ISO/IEC 27001 was published; this revised edition adopted the harmonized structure that applies to all management system standards (ISO/IEC Directives Part 1 Annex SL) and made several technical changes, for example, major changes to the risk managements processes.
- 2022: The third edition of ISO/IEC 27001 was published; this revised edition updated Annex A to align with the third edition of ISO/IEC 27002 published in 2022. Also, an amendment was added to introduce a requirement on climate control actions.

2.2.2 ISO/IEC 27001 and Related Standards Series

This section presents details of the current series of SMS standards:

2.2.2.1 ISMS Requirements

- ISO/IEC 27001: Information security management system requirements. This is the core standard in the ISO/IEC 27000 series. This standard specifies the requirements for establishing, implementing, deploying, monitoring, reviewing, maintaining, updating and improving ISMS.

2.2.2.2 ISMS Supporting Guidelines

This group of standards in the ISMS series provides advice, guidance and implementation to support the purposes of meeting the requirements in ISO/IEC 27001:

- ISO/IEC 27003: ISMS guidance;
- ISO/IEC 27004: Information security management: monitoring, measurement, analysis and evaluation;
- ISO/IEC 27005: Guidance on managing information security risks.

2.2.2.3 ISMS Controls

This group of standards covers those defined in ISO/IEC 27002 but also specific applications of the ISO/IEC 27002 controls:

- ISO/IEC 27002: Information security controls;
- ISO/IEC 27011: Information security controls based on ISO/IEC 27002 for telecommunications organizations;
- ISO/IEC 27017: Information security controls based on ISO/IEC 27002 for cloud services;
- ISO/IEC 27018: Code of practice for protection of PII in public clouds acting as PII processors;
- ISO/IEC 27019: Information security controls for the energy utility industry.

2.2.2.4 ISMS Accredited Certification and Auditing Standards

This group of standards covers requirements for third-party accredited ISMS certification audits (ISO/IEC 27006-1), audit guidelines covering first-party, second-party and third-party audits (ISO/IEC 27007) and a technical report giving guidance for assessing the implementation of ISMS controls selected through a risk-based approach for information security management (ISO/IEC 27008):

- ISO/IEC 27006-1: Requirements for bodies providing audit and certification of information security management systems—Part 1: General;
- ISO/IEC 27007: Guidelines for information security management systems auditing;
- ISO/IEC 27008: Guidelines for the assessment of information security controls.

2.2.2.5 Other ISMS-Related Standards

This final group covers a miscellaneous set of subjects: a standard that covers an overview of the series (ISO/IEC 27000). A standard that considers the governance aspects of information security (ISO/IEC 27014) through which organizations direct and control the ISMS process as specified in ISO/IEC 27001 and a technical report that covers the important topic of economics of information security (ISO/IEC 27016), this provides guidelines on how an organization can make decisions to protect information and understand the economic consequences of these decisions in the context of competing requirements for resources.

- ISO/IEC 27000: Information security management systems—overview;
- ISO/IEC 27010: Information security management for inter-sector and inter-organizational communications;
- ISO/IEC 27013: Guidance on the integrated implementation of ISO/IEC 27001 and ISO/IEC 20000-1;
- ISO/IEC 27014: Governance of information security;
- ISO/IEC 27016: Information security management—organizational economics;
- ISO/IEC 27021: Competence requirements information security management professionals;
- ISO/IEC 27022: Guidance on information security management system processes;
- ISO/IEC 27024: Government and regulatory use of ISO/IEC 27001, ISO/IEC 27002 and other information security standards;
- ISO/IEC 27028: Guidance on ISO/IEC 27002 attributes.

2.2.3 Standards Related to the ISMS Series

There are a number of additional standards that are not included in the series listed in Section 2.2.2; they are nevertheless related to the ISO/IEC 27001 series. Many of these standards provide more detailed implementation information at the control and services level. This set of standards includes the following

2.2.3.1 ICT Readiness, Information Security Incident Management, IDPS, Digital Investigations

- ISO/IEC 27031: Information and communication technology readiness for business continuity;
- ISO/IEC 27035: Information security incident management;
  - Part 1: Principles and process;
  - Part 2: Guidelines to plan and prepare for incident response;
  - Part 3: Guidelines for ICT incident response operations;
  - Part 4: Coordination.
- ISO/IEC 27037: Guidelines for the identification, collection, acquisition and preservation of digital evidence;
- ISO/IEC 27039: Selection, deployment and operation of intrusion detection and prevention systems (IDPS);
- ISO/IEC 27041: Guidance on assuring suitability and adequacy of incident investigative method;
- ISO/IEC 27042: Guidelines for the analysis and interpretation of digital evidence;
- ISO/IEC 27043: Incident investigation principles and processes;
- ISO/IEC 27050: Electronic discovery:
  - Part 1: Overview and concepts;
  - Part 2: Guidance for governance and management of electronic discovery;
  - Part 3: Code of practice for electronic discovery;
  - Part 4: Technical readiness.

2.2.3.2 Applications and Services

- ISO/IEC 27032: Guidelines for internet security;
- ISO/IEC 27033: Network security:
  - Part 1: Overview and concepts;
  - Part 2: Guidelines for the design and implementation of network security;
  - Part 3: Reference networking scenarios—threats, design techniques and control issues;
  - Part 4: Securing communications between networks using security gateways;

- Part 5: Securing communications across networks using virtual private networks (VPNs);
- Part 6: Securing wireless IP network access;
- Part 7: Guidelines for network virtualization security.

- ISO/IEC 27034: Application security:
  - Part 1: Overview and concepts;
  - Part 2: Organization normative framework;
  - Part 3: Application security management process;
  - Part 5: Protocols and application security controls data structure;
  - Part 5-1: Protocols and application security controls data structure: XML schemas;
  - Part 6: Case studies;
  - Part 7: Assurance prediction framework.
- ISO/IEC 27036: Information security for supplier relationships:
  - Part 1: Overview and concepts;
  - Part 2: Requirements;
  - Part 3: Guidelines for hardware, software and services supply chain security;
  - Part 4: Guidelines for security of cloud services.
- ISO/IEC 27040: Storage security.

2.3 Short History of ISO/IEC 27001

2.3.1 Emergence of ISMS

The third edition of ISO/IEC 27001 was published in 2022, and this was triggered by the revision of ISO/IEC 27002, also in 2022, and the need to update Annex A of ISO/IEC 27001. Further to this third edition, an amendment was published in 2024 to include a requirement on climate control actions.

Today, we have almost 30 years of history behind ISO/IEC 27001 and its global success. This takes us back to when this standard was a British standard BS 7799-2 in 1997, before it got submitted into ISO in the early 2000s. During the years of development, the standard always had business focus firmly behind it, supporting its aims and contributing business-oriented content. This played a major part in its success from the very start, moving from an IT security world to a world of information security management and business-focused requirements, with top management leading the way.

The development of information security management standards based on ISO/IEC 27001 covers several crucial management needs and requirements, not found or addressed in previous IT security standards. The past lack of management support in IT security standards left business unprotected and vulnerable. The arrival of ISO/IEC 27001 clearly filled the gap at the international level and gave organizations the chance to fully protect their business information assets, minimize their risks and maximize their investment in information security.

Managing information security using ISO/IEC 27001 replaces the old IT security view and approach, which has always been insufficient for top management and their need for protection. ISO/IEC 27001 takes organizations forward providing a tool and support for top management to realize an effective risk management-based protection of their business.

This movement toward management led standards grew out of the way as more organizations were including information security on their business agenda and having boardroom discussions on business risk and hence information security risks. All of this moved management beyond IT security, to address what was impacting the very core of business. So, the 1990s onwards were a time of change, a time to embrace these management needs and requirements.

Hence, managing your information security was focused on a people, policy, procedural and business process point of view and this is the core objective behind the approach taken in the ISO/IEC 27000 series of ISMS standards. Although we are highly dependent on technology and the role it plays in today's business environment as our information-processing workhorse, this is just one aspect and IT security is not the most significant challenge to be faced. People themselves probably present the major problem to information security, as they are the greatest of all vulnerabilities and the management of people and the processing of information present the greatest of challenges. People are risk takers when it comes to a range of everyday tasks in which they are involved, and at the same time they are risk-averse in other tasks they perform. Training, investing and motivating people, as well as allocating responsibilities for security and making them feel that they are part of a security culture, are some of the key parts of establishing an effective awareness and management system that will help businesses to protect their information. Recognition of the significance that people, procedures and process were the greatest problems to resolve in the domain of information security started to grow in the late 1980s but did not become a serious topic of standardization until the mid-1990s and today it has become an universal agenda item on management review and board level meetings.

2.3.2 Internationalization

Before 2000, the BS 7799 standards were already being used worldwide by many different industries and businesses and so they became de facto international standards from an industry perspective. The next stage in the development of these standards was to formalize them as international standards. This led in 2000 to the proposed introduction of BS 7799 Part 1 into ISO/IEC JTC 1.[1] The proposed introduction into ISO/IEC achieved the minimum majority support (67%) to be approved. Because not all member countries of JTC 1 gave a vote of approval, and it is always good to achieve a consensus, and it was decided after some debate that Part 1 would be published under the condition that an early revision of the standard should commence as soon as possible. So, in 2000, BS 7799 Part 1 became ISO/IEC 17799:2000.

In 2005 a revised version of ISO/IEC 17799:2000 was published and in the same year BS 7799 Part 2 became ISO/IEC 27001: 2005. Also, in the same year, SC 27/WG 1 adopted the ISO/IEC 27000 numbering scheme and so the "ISO/IEC 27000 series of ISMS standards" was adopted. In 2013, a second edition of ISO/IEC 27001 was published, and, in 2022, a third edition was published.

Today the international community has now adopted ISO/IEC 27001 as the "common language" for information security management. This enables organizations worldwide to engage in securing their business using such a language, in demonstrating to their customers and business partners that are "fit for purpose" to handle information in a secure way whether it be online business or off-line business. The world is now opening up to this notion of a "common ISMS language" for the benefit of all organizations to manage their risks and to protect their critical assets.

2.4 Overview of ISO/IEC 27001: 2022

2.4.1 Introduction

The international standard ISO/IEC 27001 is a set of requirements for establishing, implementing, deploying, monitoring, reviewing, maintaining, updating and improving an ISMS with respect to an organization's overall business risks and opportunities.

It belongs to a class of standards referred to as the Management System Standards (MSS), which includes standards such as ISO 9001 (Quality Management System), ISO 14001 (Environmental Management System), ISO 22000 (Food Safety Management System), ISO/IEC 20000-1 service management and ISO 22301 (Business Continuity Management System).

1. JTCI (Joint Technical Committee 1) responsible for IT-related standards.

In 2012, ISO published a harmonized approach (ISO IEC Directives Annex SL, Appendix 3) for both the development of new MSS and for the revision of existing MSS. The reasons for this were to enable an organization to operate an integrated MSS that will comply with the requirements of two or more MSS. For example, an organization could decide to operate with both ISO/IEC 27001 and ISO 22301, which would mean they would integrate their ISMS with the BCMS. Using this common "integrated" approach would provide many business benefits and increase the value of using MSS. Both ISO/IEC 27001 and ISO 22301 are two standards from the MSS family of standards that have been revised to take on this common approach, which involves using a high-level structure, identical core text and common terms and core definitions.

There have been three editions of ISO/IEC 27001: 2005 (first), 2013 (second) and 2022 (third). The second and third editions of ISO/IEC 27001 version take into account the new harmonized MSS approach. This means that the high-level structure of the chapters, clauses and sections looks different from the 2005 edition. In addition to changing the high-level structure, changes were made to the requirements specified in the standard. These changes reflected the contributions received from member bodies of SC 27 and their co-operating organizations.

This section provides some of the highlights of the third edition of ISO/IEC 27001. More specific discussion on the third edition can be found in Chapters 4 and 5.

2.4.2 ISMS Audience

The target audience of ISO/IEC 27001 did not change in the 2022 edition and the pre-international version BS 7799-2: it is suitable for all types and sizes of organization. It can be applied to any type of business activity and across all business markets, as its subject matter is the protection of information, regardless of what systems, processes or IT the organization deploys.

ISO/IEC 27001 place great emphasis on the role of top management, leadership and commitment of management in supporting the ISMS. There is emphasis on the need to align the development of the ISMS with the needs and expectations of stakeholders and all relevant interested parties and to make sure all internal and external business issues and requirements are covered.

2.4.3 Mandatory Statements of Requirement

ISO/IEC 27001 uses the word "shall" in specifying the requirements and in ISO terminology any statement of requirement, which includes this word, is mandatory to implement if an organization wishes to claim conformance with the standard. For example, "Top management shall establish (shall ensure, shall assign, shall demonstrate)—" or "The organization shall establish, implement, maintain—" are all mandatory statements of requirement that need to be actioned to claim conformance.

This means that this standard can be used for conformity assessment purposes, that is, in particular, formal third-party certification, which is similar to the ISO 9001 case for quality management systems and ISO 14001 for environmental management systems.

2.4.4 Processes

ISO/IEC 27001: 2005 was based on a Plan-Do-Check-Act process model. In the 2013 and 2022 editions of ISO/IEC 27001, this model has been excluded, although its continual improvement philosophy is certainly still firmly in place. However, the process-based approach is still very much alive and fundamental to the management system design. For example, the organization needs to have a risk assessment process to be implemented, risk treatment process, internal audit process and management review process. ISMS processes are the systematic operations and activities that are a central feature of ISO/IEC 27001.

2.4.5 ISMS Stages

The ISMS stages are establishing, implementing, deploying, monitoring, reviewing, maintaining, updating and improving and the ISMS, which the organization needs to go through a number of staged activities. These stages include "shall" requirements (mandatory requirements) where things need to be done, activities to be carried out and processes implemented. These requirements fall under the following clause headings of ISO/IEC 27001:

- The context of the organization (Clause 4);
- Leadership (Clause 5);
- Planning (Clause 6);
- Support (Clause 7);
- Operation (Clause 8);
- Performance evaluation (Clause 9);
- Improvement (Clause 10).

2.4.6 Organizational and Top Management Focus

The requirements (the "shall" statements) in ISO/IEC 27001 are all aimed at the organization and top management. The design and focus of the ISMS should be fully oriented towards the organization and its management. It is not an IT security standard; it is not driven by technology or products. It is a standard where the accountability for information security is in the hands of the organization and its top management. It is top management that needs to justify its actions and decisions regarding information security on behalf of the organization.

2.4.7 Support

Any ISO/IEC 27001 ISMS implementation needs management support to ensure there are enough resources to have an effective system of protection. This includes support for a workforce that is aware and competent to carry out their job function to ensure the ISMS performs meets the information security needs of the organization and is able to manage the risks appropriately. This support function also needs the organization to have an effective means of communicating all that is important regarding the ISMS: procedures and processes for dealing with communications in particular the what, when, with whom and how. The management support function is an ongoing activity that supplies the necessary resources to keep the ISMS operational, for making sure the risks are continually managed effectively, and that the performance of the ISMS is continuously monitored and reviewed and improvements are made to keep the ISMS up-to-date.

2.4.8 Risk-Based Approach

ISO/IEC 27001 (Clause 6) risk-based approach is critical to the organization's business activities regarding information security management and protecting their information assets. The ISMS risk management process needs to take account the requirements and expectations of all interested parties, including customers, consumers and business partners. It needs to take account of any issues that might be relevant to information security risks, be they related to corporate governance, legal, regulatory and contractual obligations, business objectives and strategy, business operations and processes or the use and application of information communications technology (ICT) and operational technology (OT) systems.

The overall risk philosophy behind ISO/IEC 27001 is all encompassing, addressing all risk-related issues that could affect the protection of information assets of the organization. Chapter 4 goes into a more detailed discussion about the risk management process, especially the assessment and treatment processes.

2.4.9 Operations

Implementation and operational use of the ISMS to meet the requirements of the organization are management functions that involves an integrated process approach. The flow of requirements taking account Clauses 4, 6, 9 and 10 across the organization needs management control of the processes involved. Planning and execution need to be carried out with the necessary management support and commitment to ensure an effective ISMS can be used operationally with confidence. Changes affecting the organization need actions taken to ensure the operational use and execution of the ISMS is managing the risks.

Regular reviews and assessment of the risks, and their treatment, are important aspects of ISMS operations.

2.4.10 Performance Evaluation and Continual Improvement

In ISO/IEC 27001 (Clause 9), "performance evaluation" is implemented through the use of several processes including taking measurements, monitoring, internal audits and management reviews. This aspect of ISO/IEC 27001 is crucial to ensure the continuing effectiveness, suitability and applicability of the ISMS for protecting the organization and its assets. Performance evaluation provides management with the vital input to decide what improvements and changes need to be made.

2.4.11 Annex A

Annex A of ISO/IEC 27001 is a reference set of information security controls based on the standard ISO/IEC 27002. It is important to note that Annex A is not a set of requirements, and its purpose is simply to support the implementation of the requirement in Clause 6 (6.1.3.c) of the standard. It is a common mistake that is often made to think that Annex A is a set of control requirements that users can go to for selecting which controls to implement. This is a wrong and incorrect understanding of Annex A and its role and purpose in ISO/IEC 27001.

2.5 Conformance to ISO/IEC 27001

The term "conformance" is often an overused or misused word. However, when it comes to international standards, in particular, management system standards such as ISO/IEC 27001, the definition is clear. Conformity assessment is defined as a demonstration that specified requirements are fulfilled. In terms of ISO/IEC 27001 it is a demonstration that all the requirements (all "shall" statements) specified in Clauses 4 through 10 of ISO/IEC 27001 of the standard are fulfilled to claim conformance. Conformity assessment can be carried out by a first, second or third party. First-party assessments include internal ISMS audits (see Clause 9.2 ISO/IEC 27001). Third-party assessments are independent certification audits.

Conformity assessment, independent certification assessments, plays an important role in building trust, confidence and assurance in an organization's ISMS implementation. It provides benefits and assurance to the stakeholders, customers, consumers and regulatory bodies involved with the organization that the ISMS is meeting their needs and expectations and is providing effective, suitable and appropriate protection. It has become an international benchmark for trust and assurance in ISMS implementations.

CHAPTER

3

Contents

3.1 ISMS Business Case

3.2 Organizational Context

3.3 Needs and Expectations

3.4 ISMS Scope

ISMS Business Case and Context

Without a purpose, nothing should be done.
—Accredited to Marcus Aurelius

3.1 ISMS Business Case

It is important to have an ISMS business case so that the necessary commitment, funding and resources can be allocated to support the design, development, implementation and continual maintenance of the ISMS. The process and design of the business case can follow that of other business cases the organization has to support other business projects it has. The business case should at least cover the consideration of the following:

(a) An understanding of the organization, its needs and requirements, risks and compliance requirements (legal, regulatory and contractual obligations) contracts;

(b) Business assets to be protected (information, people, property, facilities);

(c) Existing vulnerabilities linked to business assets, costs of incidents and disruptions versus cost of controls to mitigate the risks;

(d) Operational and commercial priorities (supporting the business, supply chains);

(e) Alignment of ISMS investment to business objectives, value for money, return on investment.

Top management and leadership within the organization must consider investment in the well-being arising from the implementation of an ISMS. Recovery and remediation of incidents can be expensive and, in some of the worst scenarios, more expensive to deal with than the cost of prevention of incidents. The cost of implementing, monitoring and maintaining an ISMS comes with a cost, but this needs to be set aside from the costs resulting from a lack of protection of business assets, lost information, damaged infrastructure and the organization's reputation.

Developing the business cases will require different types of information from different sources, such as that discussed in Sections 3.2 to 3.3 (Clause 4 of ISO/IEC 27001) and other chapters of this book. Understanding the organization and its context and the needs and expectations of interested parties is an important requirement ("shall" statements); see (a) to (e).

3.2 Organizational Context

3.2.1 Understanding the Business

Before embarking on the design, development and implementation of an ISMS, it is important to understand what the ISMS needs to cover regarding your organization, its context and purpose, its organizational structure and how it works and operates. This includes consideration of the issues, both internal and external, that are relevant to the organization's purpose and business objectives and that could influence the ISMS outcomes that the organization intends to achieve. An important factor in this respect is the issues that could pose an information security risk to the organization's business objectives, purpose and information and information systems. It is also important to understand the organization's risk culture and appetite, in order to properly design, implement and integrate the ISMS within the organization.

For the ISMS to be effective, appropriate and suitable to meet the organization's objectives and purpose, it needs to determine the risks and opportunities. Achieving an accurate measure of what these risks are and all relevant internal and external issues and the requirement to be addressed needs to be identified and considered (see Section 3.3). This means for the ISMS to effectively manage these risks; its design and implementation shall take account of these issues and relevant information security requirements and be at one, in harmony and working together, in the business context and environment of the organization.

Understanding the business and its context is essential to enable the ISMS to be an integrated part of the organization and not function as a separate, stand-alone entity. This is important because the ISMS should be a business enabler, protecting and adding value to the business, minimizing the information security risks to help maximize its business opportunities. This is why Clause 4 of ISO/IEC 27001 is the starting point in the development of an ISMS and a mandatory step for all organizations to go through.

3.2.2 Internal Issues and Context

All those internal issues and dependencies relevant to the purpose and objectives of the organization achieving an effective ISMS shall be identified and considered. For example, the organization needs to ask: "Are there any internal standards, policies and procedures related to business processes or the management of operations and resources that are relevant?" There may be issues, restrictions and dependencies that the ISMS need to take into account when considering the information security risks and the implementation of controls for treating these risks. For example, restrictions on the type of technology that can be deployed because of internal procurement policy or a dependency on another department or business unit for specific support or resources.

There may be a number of internal information systems and different protection needs for each system, protection for the sharing of information or internal communications of information. There may be IT infrastructure issues that may impact ISMS performance and effectiveness; again, these issues need to be considered in the ISMS risk assessment and treatment. The organization may operate with internal service contracts between business units and these may specify service delivery and availability issues and dependencies that the ISMS would need to address. There may be specific issues related to the internal workforce regarding organization culture, dependencies on capabilities and competence of individuals for specific roles and awareness issues.

Understanding the internal environment is especially relevant in order to appropriately address the ISMS risks. There are internal issues and dependencies that have an impact on the implementation of the ISMS, but the ISMS itself will also have a direct impact on business operations. Therefore, the better informed we are of what issues and dependencies are involved in the organization's working environment and their effect on implementing an ISMS, the better the risk assessment and the better the decision-making will be at arriving at the most effective way to treat the risks. Good risk decision-making, the type and level of risk control needed and the cost and benefits of implementing these controls depend on good information being available and duly considered during the risk management process. This also applies to the external business environment in which the organization operates.

3.2.3 External Issues and Context

All those external issues and dependencies relevant to the purpose and objectives of the organization's ISMS need also to be identified and taken into consideration. This includes the following areas of risk and business dependency:

- External organizational relationships;
    - Customers, clients, consumers;
    - Business partners;
    - Supply chains;
    - Service providers;
    - Regulators;
    - IT vendors.
- External business processes;
- External resources:
    - Connections;
    - External workforce;
    - Infrastructure elements.
- Market conditions and competition;
- New laws and regulations;
- Environment and infrastructure;
- Climate change actions and requirements and its relevance to the business.

Dependency on outside resources, services and infrastructure can all have a negative impact on the organization if, for example, there is an incident that causes disruption to the supply of these resources and services or results in damage to the infrastructure. Failures or interruptions, for

Case Study 3.1: Supply Chains

For those organizations deploying supply chains as part of their business operations, there are several issues and dependencies related to managing the information security risks. This is of specific importance to those organizations that deploy supply chains, where one or more ISMS implementations are part of the chain. For example, in any supply chain scenario, the organization is dependent on several business activities and functions being integrated with the supply chain processes. Supply chain management needs to take into consideration the security aspects for the delivery of goods or services across the chain. The business systems in the chain need to form secure partnerships to ensure they can react and respond appropriately to attacks and incidents targeted at elements in the supply chain. Therefore, it is critical that all weaknesses and vulnerabilities in the supply chain are identified. Also, a good knowledge of the issues involved is relevant to the risks and impacts to the chain.

Disruption in the supply chain can have a ripple effect across the entire chain, interrupting activities and processes that depend on each other and finally impacting the business of the end-user organization. A security disaster, incident, compromise or failure in one part of the chain could have an impact on one or more other parts in the chain, subsequently affecting one of several organizations in the chain. For example, a botnet attack or malware attack could ripple through the chain and have disastrous operational and economic effects. The incident that caused the disruption in the chain may not be man-made or IT-related, but it might have been caused by an environmental problem, for example, a flood, tsunami or earthquake. One of the factors to be considered in this instance is whether organizations in the supply chain are geographically located in high-risk areas or locations prone to natural disasters.

It is therefore essential for organizations that they look at both the business risks and the opportunities of deploying supply chains. They need to make sure they have factored in all the supply chain dependencies and issues that have a direct impact on the ISMS into their information security risk assessment, and, as a result of this assessment, they have determined an effective set of information security controls to manage these risks. They need to consider the issues of contingency, continuity and availability of services during worse-case scenarios, be prepared for the unexpected and be able to adapt and build in reliance mechanisms into their business systems and processes.

example, in energy supply, telecommunication services, transportation services and logistics services, can have a major impact on organizations.

Those organizations supplying such resources and services are responsible for managing their own information and IT security risks. However, it is the responsibility of the organization availing itself of such external resources and services to identify external risks in order to be able to appropriately manage their own business risks. This is why it is important to understand the organization's activities, operations and processes and to identify the external issues and dependencies to be able to properly manage its risks. The more well-informed the organization is about these issues, the better able it is to make decisions of how to treat and manage its risks due to external dependencies.

3.3 Needs and Expectations

3.3.1 Interested Parties

Generally, an ISMS interested party is any person, persons or organizations who may be affected by or involved in the development, implementation and operation of an organization's ISMS. Interested parties include employees of the organization managing the ISMS, investors, business partners, suppliers, customers and other interested parties such as regulatory bodies.

All employees are interested parties of the ISMS in the general sense as they all have a part to play in the implementation and management of risks related to their own job function, and their employment is dependent on the organization being commercially sound and the business doing well and being properly protected. However, some employees will have more involvement than others depending on their job function, for example, a security officer, but regardless of this, all should have a vested interest in the organization being protected.

Implementing an ISMS should give confidence to all interested parties that the risks to the confidentiality, integrity and availability of the organization's information is being properly and effectively managed.

3.3.2 Requirements Relevant to the ISMS

Once the interested parties have been identified, then the needs, expectations and requirements of these parties relevant to information security should be identified.

3.3.2.1 Stakeholders and Investors

Stakeholders and investors want a good return on their investment from the organization and so the funding and support of the ISMS are considerations.

This means that they want to see that the organization has an ISMS that is effectively managing the risks and opportunities. Stakeholders are clearly interested in maximizing the business opportunities to be achieved by implementing the ISMS and at the same time minimizing the information security risks and controlling any potential negative business impacts, as all this can affect their investment. Key to this is consideration of the requirements and expectations of interested parties in order to be able to manage the information security risks and to provide adequate protection of the organization's information and its operations to avoid any negative impact to the business.

3.3.2.2 Customers and Business Partners

Customers, clients and consumers want some confidence that their information is being protected. For example, sensitive or critical information that is shared for business reasons will be kept safe and secure and will not be leaked, stolen or accessed by those who are not authorized to see or use it. This also applies to PII that is shared or given as part of a business activity.

Business partners want some assurance that the organization is fulfilling its contractual obligations, for example, to deliver product or service as agreed, or that its commercial information is in safe hands and has not been leaked or shared with competitors. Consumers also want assurance that their information is protected, for example, in accessing services, for online buying and banking.

3.3.2.3 Connectivity and Outsourcing/Cloud

External customers or partners want some assurance that networked connections they have with the organization will not be a channel for malware, hackers or other security breaches. This means that the organization needs to fully address the technical requirements of information security related to their connections and connectivity, use of outsourcing and cloud services. There is a need to consider their policies and procedures for information exchange and sharing with external parties. They also need to consider what they have in place for incident management and what reporting agreements on incidents, which is especially relevant to supply chain management arrangements, outsourcing and cloud services with external parties.

Many organizations are outsourcing their business activities and so are dependent on business partnerships to execute various functional and operation aspects of their business. Examples of this include facilities management, transportation services, call centres and back-office activities. It also includes the increasing use of outsourcing of IT services, for example, using cloud services.

The more dependent organizations are on the use of external services; the more risks involved, the greater the potential impact on operations. The organization is more reliant on the external service provider managing its own risk.

3.3.2.4 Internal Technology Requirements

The organization may have a number of internal business requirements regarding the use of IT. For example, the organization may have particular business reasons for needing:

- Smart mobile devices;
- Cloud services and other outsourcing facilities;
- AI, generative AI and enabled virtual assistants (VAs);
- Communication tools (e.g., for business conference calls, meetings, workshops);
- IoT and 5G;
- VPNs;
- Blockchain and distributed ledger technologies;
- BYOD (bring your own devices) at work.

Case Study 3.2: Supply Chain Requirements

For those organizations deploying supply chains as part of their business operations, managing the risks needs to consider the following problems:

- Poor supplier performance and timing issues and service delivery;
- Business interruptions/disruptions and lack of continuity plans;
- Lack of resilience and protection to cyber attacks;
- Loss, theft or leakage of information;
- Shortages of human resources;
- Financial and economic disruption;
- Dependence on global supply chains and off-shore supplies.

These technologies, and many more that will emerge in the future, all have information security and privacy risks associated with them. However, most of them offer business opportunities for improving performance and productivity, meeting competitive changes and generally staying ahead of the game in the marketplace where changes are frequent. The balance of risk and opportunity is something the organization needs to address through an effective risk management framework, something we shall address in Chapter 4. The impact and cost of doing nothing to address the information security risks put the organization in a precarious position that can severely affect its performance, productivity, reputation and ultimate survival in the marketplace. Consequently, it is important for the organization to understand its internal technology requirement, understand and evaluate the risks involved and make the necessary investment to manage the risks.

Cloud, AI, IoT, blockchain and other technologies all offer a range of opportunities for business, but the organization needs to manage the information security risks associated with these technologies to protect themselves whilst gaining from their investment in these technologies.

One of the other aspects of the modern business environment is of course the use of mobile devices, mobile networks and services, and carrying out some aspects of business remotely. Mobile devices may be phones and tablets provided by and owned by the organization itself. Mobile devices, networks and services all have information security risks related to them and so it is important that an organization identifies the requirements for such IT capability and understands and manages the risks associated with such technology and takes account of its requirements for the use of such technology in its ISMS risk assessment. This includes dealing with issues of theft, what types of business transactions are allowed using such technology, whether remote connections to the organization's servers are allowed, use of wireless networks off-site, use of such technology in public places, introduction of malware, noncompliance with regulations, and secure disposal of such devices. In addition, more use is being made of cloud services and again the organization needs to carefully consider its reasons for using such services, which can provide many business benefits, but the security aspects need also to be considered. Organizations are finding cloud services such as the use of software application services, data storage and management and website hosting to be economically and operational beneficial. However, the information security risks raises questions that the organization needs to understand and be able to protect itself, this includes the security provided by third parties to safeguarding sensitive and commercial information, as well as any PII that is in the cloud, the compliance issues relating to the protection of PII, data segregation of the organization's information with that of other organizations, physical location of the servers used to deliver the cloud services, data recovery in the case of a major disaster at the cloud server sites and investigating incidents at the server sites.

Some organizations allow their employees to use BYOD at work. Such a requirement means there are a number of information security issues to be considered as to what is acceptable use of BYOD on-site and off-site (at home and in public places) as well as the security of BYOD itself. The use of BYOD raises questions of protection of business information on a device that contains private information, compliance issues, technical support issues, incompatibility with other IT used by the organization, introduction of malicious code onto the organization's network via BYOD and other risks.

Another area that the organization needs to consider is its strategy and policy, and hence its requirements, regarding the use of social networking/media, by both the business and employees. There are many risks involved with the use of social networking/media sites that both the organization and its employees need to be aware of. The organization therefore needs to take account of its requirements for the use of social networking/media in its ISMS risk assessment. Its requirements need to be considered alongside the potential risks: accidental or innocent disclosure of company confidential information by employees, intentional disclosure of such information for the purpose of fraud, financial gain or identity theft and the disgruntled employee who wishes to damage the reputation of the organization. Other risks include online abuse, stalking, access to and downloading of inappropriate material, social engineering attacks (persuading employees to give information away or tricking them into doing something they should not be doing) and the introduction of malware embedded in attachments.

The organization will need to consider its requirements of the issue of acceptable use of all its IT resources and facilities by employees, for example, the use of BYOD, mobile devices and networks, cloud services and social media sites, which can be included in a corporate Acceptable Use Policy.

3.3.3 Gathering Requirements Relevant to the ISMS

Information on the needs and expectations of interested parties for information security can be gathered in various ways and from different sources:

- Internal business objectives and requirements:
    - Corporate policy and objectives;
    - Heads of business units and departments, project owners, employees;
    - Risk owners.

- Requirements specified in contracts and service level agreements (SLAs):
    - Clients and customers will want the organization to comply with any clauses dealing with information and IT security included in contracts and SLAs; for example, this may be a requirement for 24 × 7 data management services, call centre services and supply chains.
    - A client may specify, in a contract, specific requirements for reporting on incidents that may affect them.
- Compliance requirements defined in legislation and regulations:
    - There are many laws on the protection of personal data that detail specific information security requirements.
    - There are telecommunications regulations that specify information security requirements.
    - There are laws that relate to the admissibility of electronic evidence that organizations should be aware of regarding the collection of evidence during a security incident.
    - Requirements related to climate control actions.

Gathering information on the needs and expectations of interested parties can be done in different ways:

- Meetings and interviews with internal and external interested parties;
- Internal management review meetings;
- Feedback from clients, customers, business partners and employees.

3.4 ISMS Scope

3.4.1 What to Consider and What to Include

Defining the ISMS scope is an important initial task in development of the ISMS. It is important to get this aspect right before the organization spends time and effort in carrying out a risk assessment. This might sound an obvious thing to do, but defining the ISMS scope can present a number of challenges.

The scope needs to take account of the internal and external issues referred to in Section 3.1 (ISO/IEC 27001 Clause 4.1) and the requirements referred to in Section 3.2 (ISO/IEC 27001 Clause 4.2). In addition, the organization needs to consider all relevant interfaces and dependencies between its business operations and those of other organizations.

The ISMS scope can cover all or part of the organization, so we need to define its boundary in the case that the ISMS scope is only part of the organization. The extent and size of the ISMS scope, whether it is all or just part of the organization, are management decisions and depend on the organization's business aims and objectives and the motivation for its particular choice of ISMS scope.

The ISMS scope might cover the whole organization (Company A) or part of the organization (Company B) or an organization might choose several ISMS implementations (Company C). More precisely, for Company C, the ISMS scope might be defined and cover the activities of a business unit or department, a project activity or a set of business functions or services (Figure 3.1).

3.4.2 Object of ISMS Scope

Experience and feedback of existing implementations around the world inform us that organizations are selecting the ISMS route and a particular scope of application to achieve at least one of the following aims:

- Minimize the business impact and financial costs of security incidents in one or more business areas to:
    - Improve service delivery to customers;
    - Increase in staff morale, productivity, effectiveness;
    - Reduce loss of contracts, sales, orders or profits;
    - Improve and enhance customer trust, confidence and relations, and meeting contractual obligations of its customer(s), such as in the case of a company providing managed services to its clients;
    - Improve market position;
    - Reduce loss of assets and asset value;
    - Reduce risks;

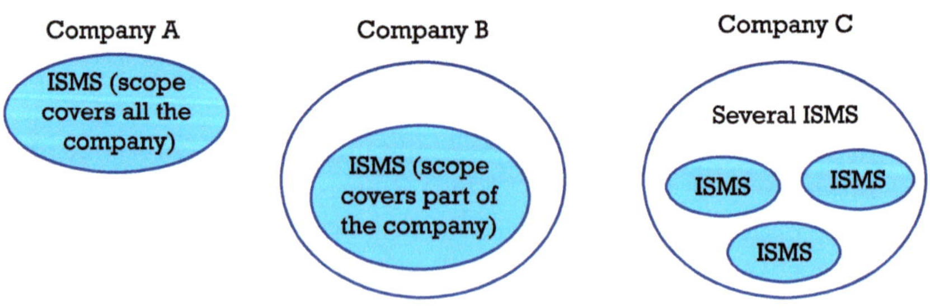

Figure 3.1 ISMS scope scenarios.

- Avoid legal/contractual penalties/liabilities;
- Reduce breaches of operating controls;
- Reduce the number and frequency of incidents in one or more business areas:
  - Human errors;
  - Service interruptions;
  - Misuse/abuse of company resources;
  - Theft, fraud and other related crimes;
  - Malfunctions, system failures, system downtimes.
- Complying with governance rules and a set of laws and regulations, such as the cyber-legislation in different parts of the world, data privacy directive in the European Union (EU) and the equivalent laws in the EU Member States, or equivalent laws in non-European countries, or for complying with the Sarbanes Oxley (SoX) or HIPPA regulations in the United States;
- As a market differentiator, for example, a company that offers online (e.g., internet banking service) or off-line (e.g., media disposal, data recovery) services;
- Enhancing corporate value of their business and as a business enabler;
  - Maximizing business opportunities and investments;
  - Providing more informed decision-making about security risks;
  - Improving risk awareness, risk control and information security effectiveness for the organization's sensitive and critical assets;
  - Facilitating an organizational information security risk prevention culture;
- Public demonstration that they are "fit for purpose" by means of a third-party independent audit, such as companies providing a particular line or type of customer facing services (help desk and call centre facilities), as part of a supply chain, as a supplier of products, or as a supplier of infrastructure supporting services (electronic signatures for secure payments and transactions).

3.4.3 Defining the ISMS Scope

Whatever the organization decides its ISMS scope to be, it must ensure that it is well-defined and covers all that it should cover in terms of requirements (ISO/IEC 27001 Clause 4.2) and issues (ISO/IEC 27001 Clause 4.1). The organization should not exclude things from within the ISMS boundary if they seem to be difficult to deal with as such exclusions could affect the organization's ability, and/or responsibility, to provide information security that meets the security requirements determined by risk assessment and applicable legislative and regulatory requirements.

Depending on the type of scope, the ISMS might typically include:

- The staff and employees operationally involved in the ISMS area of work;
- The processes and services used within the ISMS area of work;
- The information and information systems necessary to carry out the business of the ISMS;
- The policies, procedures and documentation to be deployed necessary to carry out the business of the ISMS;
- The interfaces, dependencies and connections to the ISMS;
- Supporting IT infrastructure for the ISMS;
- The physical location(s) of the ISMS.

3.4.4 Scope Example

An example of ISMS scope is a sales and customer services department of a small manufacturing company. This is just one of several departments in the company as shown in Figure 3.2.

The scope of the ISMS will typically involve:

- The personnel sales and customer services;
- The procedures used by the personnel;
    - Company policies and procedures;

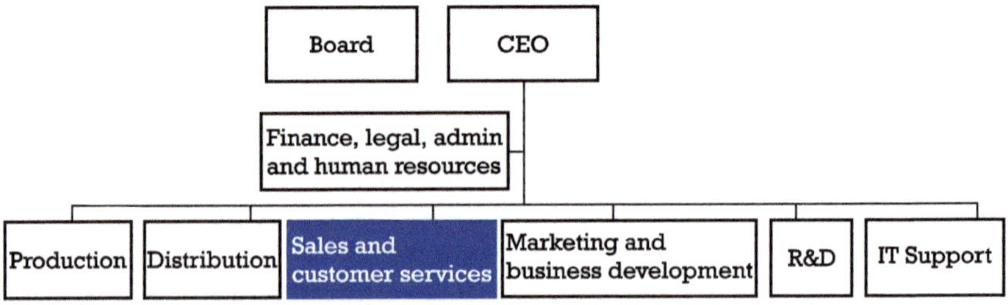

Figure 3.2 Organizational structure of a manufacturing company.

    - Operating manual;
    - Security policies and procedures;
- Sales and customer service information;
    - Customer details;
    - Customer orders, invoices;
    - Internal company data;
    - Data from other companies;
- Use of business processes, applications and services;
    - Sales orders process;
    - E-commerce;
    - Customer enquires process;
    - IT services;
- Interfaces and dependencies;
    - Technical;
    - Nontechnical;
- Use of technology;
    - Desktop computers, telephones, mobiles, tablets and other IT equipment;
    - Cloud services;
    - Website presence;
    - Networks;
- Physical location.

This example illustrates an ISMS scope that is only part of the company operations and the services that it provides. Other scoping examples include:

- Internal services supplied by the IT department;
- Call centres;
- Internal help desk support;
- An organization's claims and payments department for its customers;
- Logistics support for a shipping or airline company;
- Online hotel booking system for an international hotel chain;
- Research and development group of a software company.

3.4.5 External and Internal Connections

An important aspect of the ISMS scope is the connections that exist to and from the ISMS, be these with external customer or supplier systems or be they internal with other parts of the same organization. In the example given above, typical internal connections might be with the personnel department, those dealing with IT services and those dealing with physical security. External connections might typically be those with other banks and network and ISP service providers. It is important to consider these connection interfaces as these interfaces are one of the ways that risks might present themselves to the ISMS. Management and control of the flows of information across these interfaces are important to protect the ISMS against these risks.

Managing these interfaces can be done in many ways through the use of contracts, SLA and use of operational procedures and technical controls.

CHAPTER

4

Contents

4.1 Importance of Risk and Opportunity

4.2 Risk Management Process

4.3 Ongoing Reassessment of Risk

Managing the ISMS Risks

It's not because things are difficult that we dare not venture. It's because we dare not venture that they are difficult.

—Accredited to Seneca

4.1 Importance of Risk and Opportunity

4.1.1 Risk Definition

The definition of risk in the 2022 edition of ISO/IEC 27001 is generic to facilitate a greater range of methods and techniques for assessing risks. There are many definitions of risk as the term is used in several difference contexts. The definition of risk (defined in ISO 31000) adopted in many areas of ISO, including ISO/IEC 27001, is that "risk is the effect of uncertainty (on objectives)." The elements of this definition can be interpreted in several ways, so effect can be regarded as the consequences or impact of an event occurring such as the consequences of a malware incident on an email server.

The uncertainty could be that related to the unpredictability or chance of an event or incident occurring, for example, the chance of an unauthorized access attempt on a system containing sensitive information.

In the case of ISO/IEC 27001, the objectives are to preserve the confidentiality, integrity or availability of information. The ISMS target is to provide protection for these objectives that is effective, suitable and adequate to meet the organization's needs and requirements. For example, the business impact of an unauthorized access attempt on a system containing sensitive commercial information would be detrimental to the organization, not only damaging their reputation but also possibly suffering financial loss and potential legal penalties, and the chances or possibility of this happening might be high.

4.1.2 Opportunity

An organization needs to identify the opportunities that will contribute to achieving its strategic and business objectives. This can sometimes mean taking risks to gain access to these opportunities and sometimes reducing the risks to ensure that these opportunities are available. The effect element in the definition of risk, given in Section 4.1.1, can also be interpreted as the deviation from the expected, that is, in terms of both positive as well as negative results, consequences or outcomes. This might be also expressed in terms of good or bad fortune, exposure to gain or loss, upside risk (favourable outcome) or downside risk (unfavourable outcome). This means that an organization needs to create a balance between those negative risks that prevent opportunity and those positive risks that help to secure opportunities, and this will depend on the organization's risk attitude, tolerance and appetite.

4.1.3 Risk Attitude, Tolerance and Appetite

Should a business miss an opportunity to avoid risks or should the business accept or take risks to gain from the opportunity? This relates to the risk attitude of the organization: are they risk-averse (avoiding), risk neutral or risk-seeking (high appetite) when it comes to going for business opportunities? The risk-averse (risk-avoiding) individual will try to reduce the uncertainty of a risk, that is, when faced with either a choice where the loss or gain is uncertain or an option that is more certain but may result in lower gain, they would select the latter option, the more certain option. The risk seeker takes a gamble on the uncertainty in the hope of a greater gain.

The risk attitude is a factor in towards determining their risk appetite to a situation or set of business choices. Risk appetite is the amount and type of risk that an organization is willing to pursue, take, accept or retain to achieve its strategic and business objectives. Hence, an organization's risk appetite is linked to the pursuit of a business opportunity. Different organizations have different risk appetites. An organization with a low-risk appetite wants to avoid the negative effects of the uncertainty of risk, and they want to pursue guaranteed gain from a business investment and not gamble on an uncertainty for a much higher gain; these are risk-averse types of organizations. At the other end of the spectrum, there are those organizations that are risk seekers, and these seek to gain higher rewards by selecting business opportunities and business innovations. Risk tolerance is the level or amount of risk that an organization is willing and able to accept or not accept.

The risk appetite and the levels of risk tolerance need to be defined by top management in conjunction with the board of directors and the CEO.

4.1.4 Information Security Risk Appetite and Tolerance

With regard to information security risk appetite, this should be a part of the organization's overall risk appetite. Defining the information security risk appetite, as with other risk decisions and criteria, should not therefore be an isolated activity but an integrated part of the organization's overall strategy and objectives on risk. ISO/IEC 27001 emphasizes the importance that management of risk comes from the top, with commitment, direction and decisions being made by executive or senior management.

Information security management should be seen as a driver for achieving the organization's business strategy and objectives and for adding value to the organization. Therefore, the risks related to information security need to be managed to achieve such business objectives and to satisfy internal and external requirements. The outcome and deliverables of the ISMS are therefore to provide effective protection of the organization's information to achieve these business objectives and requirements.

Traditionally, information security risks have been seen in a negative way, to protect against loss or damage, and not in the context of business opportunity to address the broader business strategy. However, as just mentioned, information security is not an isolated activity but one of the key drivers for an organization to protect its business interests through protecting the information it uses to run and operate its business.

There are information security risks related to governance and compliance requirements, which the organization needs to address and, because of the nature of these types of requirements, sometimes there is no clear view of any underlying business opportunity. However, even in the cases of mandatory compliance, there is always an aspect of business opportunity, which is especially the case as the approach to the management of risk is taken as part of business strategy. Conformity to the requirements of ISO/IEC 27001 demonstrates that the organization is protecting its information and this includes addressing the risks using legal controls and protection. For example, the protection of personal data or PII should start with the compliance with laws and regulations pertaining to the country of business. An example of this might be any service that an organization offers, which involves managing and processing customer data including customer PII. The service makes money for the organization and so provides a business opportunity, but, in pursuit of this opportunity, the organization needs to comply the legal requirements that are relevant to its customer data and ensure adequate protection is provided in conformity with these requirements.

In considering information security appetite and tolerance, executive management decisions need to be made taking into consideration any legal advice regarding any regulatory and legislative aspects.

4.1.5 ISMS Risks

Managing the effect of uncertainty on achieving the objectives, outcomes, targets and deliverables of the ISMS is the task of risk management. Are the information security risks managed to a tolerable level of acceptance so that the ISMS is effective, suitable and adequate for the organization's needs and requirements? Does the organization have effective, suitable and adequate ISMS policies, procedures, resources, processes, plans and programmes in place associated with managing the risks?

Things can change in an organization and its business environment that could have an effect on its risk profile: the business objectives and strategy may vary or change; operating conditions might change; market conditions might fluctuate and there may be economic downturns and upturns; the conditions surrounding and causing security incidents might change, and the more uncertainty regarding these incidents, the greater threat and attack levels; changes to infrastructure may be internal and external; different business processes are adopted; and new and advanced technology is deployed by the business. For example, the organization might expand in a new business area and at the same introduce new technological solutions and deploy more staff as part of this expansion. The question to be asked is whether this expansion leads to changes that create new risks, or increase the level of current known risks, to the protection of its information. Is the ISMS still able to be effective, suitable and adequate to the organization? This is what management should be asking. When changes occur, the organization should reassess the risks and the opportunities associated with these changes as part of their continual improvement programme, that is, addressing the risks to ensure that the ISMS remains effective, suitable and adequate for the organization's needs and requirements.

4.2 Risk Management Process

4.2.1 Changes in the Process

As mentioned above, risk is the effect of uncertainty (on objectives). In the 2005 version of ISO/IEC 27001, there was the mention of assets, threats and vulnerabilities in relation to the definition risk. So there was a need to identify the threats to assets and the vulnerabilities of the assets that the threats could exploit. These elements that made up part of the definition of risk in the 2005 no longer exist in the 2013 version of the standard. However, this does not mean that an organization that uses a risk assessment method using these elements needs to change its method. The 2013 definition is generic enough to accommodate a wide range of methods. What it does mean is that there is no mandatory requirement, in the 2013 edition, for an organization to identify its assets, threats and vulnerabilities as part of its risk assessment, as was the case in the 2005 edition.

The overall process stages are the same: risk assessment and then risk treatment. Also, stages in the risk assessment process are the same: risk identification, risk analysis and risk evaluation. What is different is in the detail of what constitutes identification, analysis and evaluation, that is, in the 2005 edition assessment is in terms of assets, threats and vulnerabilities, but this is not the case in the 2013 or the 2015 editions: the 2022 and 2013 editions use the same approach to risk assessment.

4.2.2 Risk Assessment

4.2.2.1 Process

It is entirely up to the organization to define what information security risk assessment process or method it uses. This includes what criteria it uses to identify and analyse its risks, what acceptance criteria it uses to evaluate the risks it has identified and what risk scales it uses. The 2022 edition of ISO/IEC 27001 gives the organization a lot of flexibility to define its own process to suit its own business culture and to select the methods and techniques it wants to use as part of this process.

The assessment process should provide the organization with an understanding of the risks that they face. This is typically in terms of the risk causes and effects (consequences and impacts) and the likelihood of occurrence, the levels of risk and whether these levels are acceptable or nonacceptable in accordance with the acceptance criteria as defined by the organization.

4.2.2.2 Acceptance Criteria

What is acceptable or tolerable or unacceptable or intolerable is defined by the organization. The criteria for determining what are acceptable levels of risk will depend on the organization's risk appetite and the significance of the risks to the organization's business, objectives and values. How significant and important the risks are should also reflect both internal and external interests and requirements, relevant standards, laws and regulations.

4.2.2.3 Risk Identification

The process of identifying information security risks involves the identification and recognition of what might or could happen that may have an impact on the organization's ISMS and its ability to realize its business strategy and objectives. This includes events, incidents, potential breaches and compromises or any other situation that might have an impact on the organization. Another aspect of this risk assessment stage is the identification of existing ISMS controls that are in place.

The identification process involves gathering together information regarding the risks obtained from individual and team interviews; feedback, opinions and views from stakeholders, interested parties and experts: questionnaires or checklists; records of information security incident management records; and other historical data records: audit reports and various other methods and sources.

It is important that those involved in the information gathering and identification process recognize that risk perception can vary from individual to another, for example, what one individual perceives as a risk or a risk-related event may be differently perceived by another individual. Those in control of the identification process need to take into account the possible variations in risk perceptions both from a human perspective as well as corporate perspective.

Although ISO/IEC 27001 is dealing with information security risks, there are other areas of risk such as aspects of financial risk, project risks, human risks, supply chain risks or environmental risks that, whilst not the subject of ISO/IEC 27001, may have an impact on the ISMS and so may need to be identified during the organizational context phase (Chapter 4 of ISO/IEC 27001).

The output of the identification process should be a list of risks, which will then be used during the risk analysis process.

4.2.2.4 Risk Analysis

The analysis stage deals with determining the potential consequences or impacts of the risks, the likelihood of the risks occurring and the level of the risks. The basic level of risk is the combination of the consequences or impacts and the likelihoods.

> *Case Study 4.1: Example Risk Levels*
>
> The following illustrates an example risk method for determining the level of risk given the potential impact (VL = very low through VH = very high) and likelihood (VL = very low through VH = very high). The risk levels in this example are given numerical values: 1 = low risk through 9 = very high risk. Clearly, this illustrates that on a sliding scale that the higher the impact and the higher the likelihood, the higher the risk level. Note that there are other representations that can be used to present levels of risk. The one used in Figure 4.1 is a common one to illustrate the principles of risk assessment.

During this stage of risk assessment, the causes and sources of risk need to be understood and considered and subsequently the impacts and likelihood of these risks related to these causes need to be analysed. There can be complex cause-and-effect scenarios: a set of different events add up to a specific risk. For example, an untrained member of staff who has little experience uses a badly written operational procedure, and this results in sensitive information being sent as an unprotected email attachment. The staff member who has more experience and awareness of how to handle sensitive information may ignore the badly written procedure and protect the email attachment. This sequence of events and causes resulted in a risk of its sensitive information for the organization.

An event may result in multiple consequences, which combine together to have an impact on several information security objectives. It might require a combination of causes to have an impact on the organization. It is essential that the knock-on effects of causes and consequences are identified including any cascading and cumulative consequences.

Impact \ Likelihood	VL	L	M	H	VH
VH	5	6	7	8	9
H	4	5	6	7	8
M	3	4	5	6	7
L	2	3	4	5	6
VL	1	2	3	4	5

Figure 4.1 Example risk table.

There are different ways and methods of doing the risk analysis as there are doing risk identification. A group of staff carrying out a brainstorming or interviewing exercise is certainly relevant for a risk identification but not for calculating the impact or likelihood of the risks where a more systematic and analytic method is more useful and applicable. However, a cause-effect (scenario) analysis or a business impact analysis (BIA) is clearly relevant to the risk analysis with regard to the consequences or impacts of the risk. Furthermore root-cause analysis methods and what-if methods cover both the consequence or impact and likelihood aspects of risk analysis.

In carrying out the risk analysis, it is important to understand the events that are likely to have an impact of the types of information the organization has: sensitive, critical, personal, restricted, private and public information. The preservation of this information may be at risk, specifically regarding its confidentiality, integrity and availability. In addition, the protection of PII is also an issue. So analysis of the events and causes that threaten this preservation is essential. This analysis can be carried out using different methods including some of those mentioned above (e.g., cause-effect analysis, BIA), with input and feedback from various sources that are a legitimate and authorized interested party in this information (e.g., information owners, risk owners, top or senior management, users). In the same way that preserving business information is important, so are the critical business systems, process or services that are associated with the ISMS directly or indirectly. Such critical elements should be understood and their risks identified. Disruption, unavailability or continuity to these critical parts will have a high impact and detrimental consequences on the organization and could have a dire effect on the organization's operations. The complexity of the ISMS being risk-assessed will depend on the resources (cost, time, competence) needed to do the risk analysis. This in part also depends on the scope of the ISMS; a well-focused scope oriented to a single business group, department, project or service, with well-defined, noncomplex interfaces and dependencies, is one level of complexity, whereas a scope that covers many operational sites, many parts of the business, many projects or services with a complex set of interfaces and dependencies is a different, higher level of complexity.

4.2.2.5 Risk Table Scales

The organization has the choice of which risk method it wants to use, the criteria for acceptance and the scales for assigning the levels of likelihood and the types and levels of impact. In the example given in Table 4.1, low, medium, high and very high are used for levels of impact and likelihood. Instead, an organization might decide to use a 3-point or 4-point scale, or a 6-point or higher scale. The organization may wish to provide a more

Case Study 4.2: Risk Scale Descriptors

Table 4.1 shows examples to illustrate how this might look.

Table 4.1
Example of Scale Descriptors for Impact and Likelihood

Impact		Example Description	Likelihood	Example Description
Very high	Catastrophic	Threatens the survival of the company	Very high	Very likely (e.g., at least 4 in 5 (80%) chance)
High	Major	Major/significant losses, damage, harm	High	Likely (e.g., at least 1 in 2 (50%) chance or more)
Medium	Moderate	Moderate losses, damage, harm	Medium	Likely (e.g., around at least 1 in 5 (20%) chance)
Low	Minor	Minor losses, damage, harm	Low	Around 1 in 10 (10%) chance
Very low	Negligible	Little or no losses, damage, harm	Very low	Around 1 in 1000 (0.1%) chance or below

descriptive meaning to these points on the scale: low, medium, high and very high.

It is entirely up to the organization whether it adopts a set of descriptors and the content of the descriptor. Whatever is adopted by the organization should be aligned with its business objectives, culture and internal standards and methods it chooses to adopt; for example, it would make sense to adopt the same descriptors it uses for other risk management activities it undertakes to achieve a corporate consistency, but again this is entirely up to the organization to decide; the examples above are merely generic illustrations to described the overall risk assessment process.

4.2.2.6 Risk Analysis of Existing ISMS Controls

Defining the level of risk should take account of existing ISMS controls and their suitability, adequacy and effectiveness. This involves considering what existing controls there are in place and their relevance to each of the risks that are identified and whether these controls are suitable and adequate to modify the risk to a tolerable level.

What level of effectiveness is provided by these existing controls? Are they user-friendly or difficult to use? Do they work in the way they were intended to work? Are measurements available that will indicate the level of effectiveness?

In considering existing controls, one of the decision-making questions is whether it is best to improve the implementation of these existing controls to adequately modify the risks identified or whether it is best to implement a different set of controls.

4.2.2.7 Risk Evaluation

The evaluation stage compares the results of the risk analysis stage; that is, the levels of risk with the organization's acceptance criteria.

4.2.2.8 Risk Evaluation Decision-Making

At the evaluation stage of the risk assessment, decisions need to be made as to whether risk treatment is needed to modify the identified risks to an acceptable level. The risks in the middle band in our example in Figure 4.2 may, after an analysis of the business costs and benefits of risk treatment and business opportunities against the business impacts, fall into the nonacceptable region (the upper band in the dark grey area) or they may stay in the middle band and, for example, may be classed as for "close monitoring and an early review/reassessment" to make sure the situation can be managed and does not change and creep into the upper band. Therefore, in the middle band, there is likely to be a scale of risks from the boundary (lower band) of almost being acceptable to the boundary of being not acceptable (upper band) and so the cost-benefits, business opportunities and business impacts need some comparative management decision-making. One thing that is certain is that the cost of risk treatment needs to be kept in check to avoid the cost of modifying the risk becomes disproportionate to the business benefits and opportunities to be gained in engaging in a business

> ***Case Study 4.3: Risk Acceptance Heat Map***
>
> For example, in the example representation given in Figure 4.1, risk levels of 7 and above might be considered as unacceptable, intolerable (dark blue area in Figure 4.2), whereas levels 1 to 3 are considered acceptable, tolerable and, in general, negligible (white area in Figure 4.2). However, risk levels 4 to 6 might require further analysis on a case-by-case basis, taking into account the business costs and benefits of risk treatment, and business opportunities against the business impacts, to determine if there are any risks at these levels that are acceptable or unacceptable (medium blue area in Figure 4.2).
>
> Figure 4.2 is merely one example of a risk acceptance heat map. The organization needs to define its own light blue, dark blue and white areas, and this will depend on the organization's risk appetite.

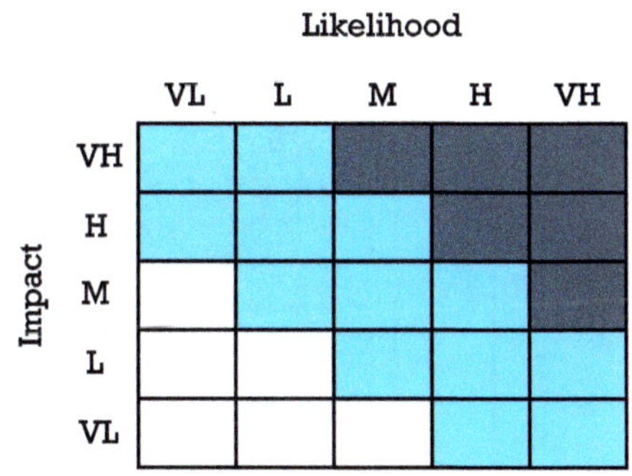

Figure 4.2 Example of risk acceptance heat map (1).

activity that is associated with the cause of the risks. Therefore, management and risk owners need to be involved in this type of decision-making where the costs and benefits of risk treatment versus nonrisk treatment, balanced out with the business opportunities and impacts are being considered.

This decision-making needs to consider the prioritization of the risks for treatment purposes. One of the major purposes of risk prioritization is to help in the allocation of resources to the treatment of risks and the implementation of ISMS controls to modify the risks and the ongoing daily maintenance and upkeep of the ISMS.

Risks should be prioritized according to the most critical through least critical ranking, for example, very high impact, very high likelihood risks. The prioritization needs to take into account an assessment of the overall impact on the organization and its critical systems and processes. High-impact risks regarding the commercial interests of the organization needs attention and generally fall into the category of critical, as might high-impact risks that might affect critical business systems or processes. The risk might be of high-impact but only medium likelihood (falling into the medium

> ***Case Study 4.4: Heat Map Examples***
>
> For example, an organization might adopt the heat map in Figure 4.3 as it has a lower-risk acceptance.
>
> Another organization might adopt the heat map in Figure 4.4 as it has a higher-risk acceptance.
>
> The choice of risk criteria acceptance and risk appetite is entirely up to each organization to define. Figures 4.2 to 4.4 are merely examples to illustrate the theory. Practical implementation is up to the organization to select which risk criteria are suitable and adequate for their own business situation and circumstances.

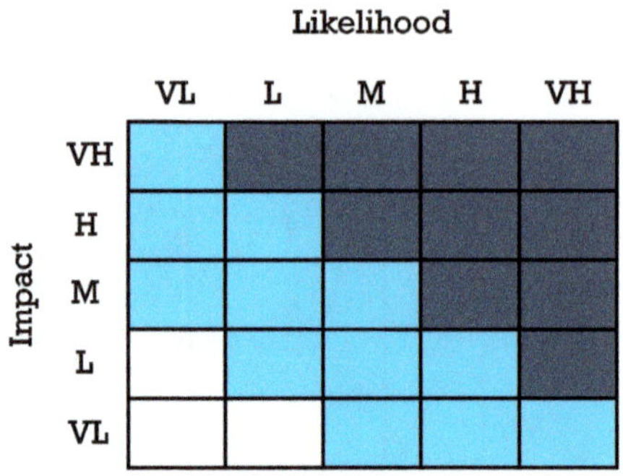

Figure 4.3 Example of a risk acceptance heat map (2).

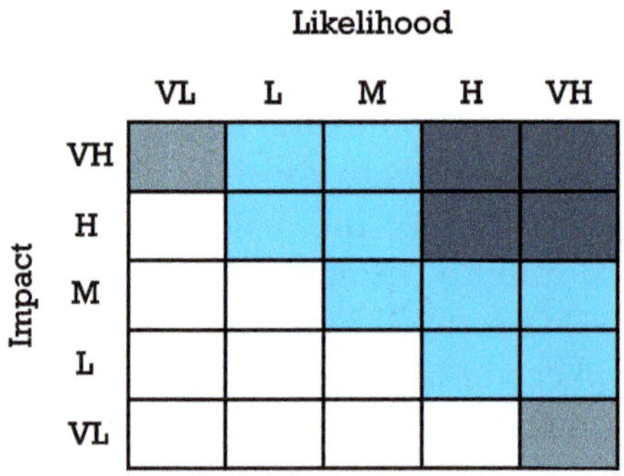

Figure 4.4 Example of a risk acceptance heat map (3).

blue area); nevertheless, the risk relates to some piece of legislation, for example, protection of personal data, in which case, the priority criteria are compliance with legislation.

4.2.3 Risk Treatment

After identifying the risks and prioritizing the risks for treatment, then the next process is to consider the treatment options that will modify the risks. The risk treatment process is typically an iterative process, as illustrated in Figure 4.5.

Also included in the process is assessing how effective the risk treatment is at modifying the risk.

The selection process needs to be a decision-making process (see Section 4.2.2) involving management as considering the costs of treatment implementation against the business benefits that come from such treatment taking account the compliance requirements from legal, regulatory

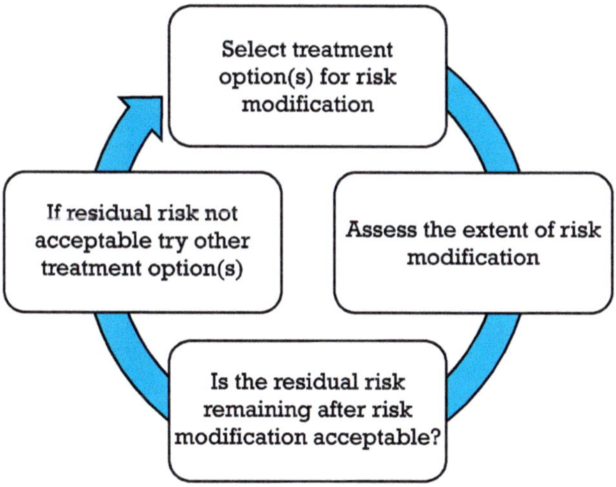

Figure 4.5 Risk treatment cycle.

and contractual obligations. It might be the case that there are strong financial and business impact reasons regarding whether a treatment option is feasible. As discussed in Section 4.2.2, it might be the case, for example, that the impact on the organization of damage to its reputation is very high, but the likelihood of this happening is considered to be low to medium, which would place the risk into the middle band of risk, neither tolerable nor intolerable. This would mean further management discussion and enquiry into the economics of treatment.

The following are some typical types of treatment options:

- Risk avoidance (risk-averse):
    - Terminate or abandon activities that currently cause or give rise to the risk.
    - Change, postpone or cancel plans, projects or activities that could cause the risk.
- Risk-taking (risk affinity): pursuit of opportunity;
- Removing the source of the risk;
- Risk transfer:
    - Insurance: passing the financial impact of the risk to an insurer;
    - Contracts: transfer by outsourcing the activity or passing the cost of loss on by contract (e.g., limitation of liability).
- Retain the risk: knowingly and objectively accepting the risks through well informed decisions;

- Modifying the risk by reducing or changing the likelihood or modifying the risk by reducing the consequences or impact;
- Risk financing: financing the cost of retained losses.

4.2.3.1 Risk Avoidance

Avoidance includes making the decision to not carry out, engage in or perform an activity that could cause the risk. Risk avoidance might involve changing:

- Company policy or strategy;
- Plans for location or relocating of business facilities;
- Methods, procedures or processes used;
- Operational conditions;
- Scope and objectives of new projects, developments or technology being used.

An example might be the relocation of the business to an area that is found to be less environmentally hazardous in order to not take on the liability that comes with the likelihood of these hazards occurring and interrupting, disrupting or destroying the business and its operations. It might be that the site for relocation is in a less dangerous area of a city to avoid serious incidents involving criminal activities and civil disorder. Avoiding this risk may seem a way of dealing with this problem but consideration needs to be given as to whether the organization will be at a commercial disadvantage or will lose out on the benefits of not relocating.

Management thinking needs to weigh up the pros and cons of either avoiding the risk in order to take the opportunity of safer operating conditions or taking the risk in order to take the opportunity of other business benefits that by be afforded by the current location such as its customer base. The same avoidance strategy might be taken regarding the risks of taking on new research projects, selecting backup sites for business continuity purposes, embracing new technologies such as digital accounting or online business.

4.2.3.2 Transferring the Risks

This treatment method involves employing another party, such as insurance company, suppliers, venture capital business and external parties, to share some or all of the risk. Typically, this could be by contract, insurance or some form of hedging. Insurance is one type of risk transfer that uses contracts and in recent times e-risks insurance protection has started to emerge on world markets. Other means of transfer might involve the use of contract language that transfers the risk to another party without the use of an insurance premium. Such a contract might set limits of liability on contractors, or service suppliers are very often transferred this way. In the financial world, there are many other instruments of transfer for financially managing risks such as taking offsetting positions in order to hedge, for example, against investment losses.

4.2.3.3 Reducing the Likelihood of the Risks

This involves, for example, using preventive, deterrent or detective types of control methods to reduce the likelihood of the risk happening in the first place or at least make it more difficult for the risk to occur.

Information security policies, procedures and processes, user training and awareness programmes, removal of weaknesses and access control methods, all these and other measures, can help to reduce the likelihood of risks. The actual effectiveness of these controls to reduce the likelihood will depend on the quality and effectiveness of the implementation and its operational or business use. For example, a control designed to prevent a security incident from being successful may not be effective in practice if the implementation is faulty or not properly done and, if users are not trained in the use of the control, this also reduces the effectiveness of the control.

Procedures for instructing users how to handle sensitive information can reduce the likelihood of errors and mistakes being made by users and the risk of such information being leaked to unauthorized users. Similarly, a process for securely disposing of no longer used media can reduce the likelihood of information stored on that media falling into the hands of those unauthorized to see such information.

Controls that are able to protect against malware incidents help to reduce the likelihood of malware infections. If those same controls have the ability to recover from malware infections, they also help in reducing the impact of the infection if they are used in an effective and timely way.

Access controls can restrict the systems, applications, privileges and information that users can access, and the stronger these controls are, the less likely it will be that unauthorized access will take place. This means that the more difficult it is to penetrate through the access control system, then the likelihood of the risk happening is lowered and much lower than if there was no access control system or the system was weak, that is, easy to penetrate.

There are many controls that have the ability to reduce the likelihood of a risk happening and being realized. When a risk does materialize, then it can have a direct impact on the operational performance of the ISMS and so controls to reduce or minimize this impact should also be considered. Some controls are able to work in combination to achieve both a reduction in likelihood and impact.

4.2.3.4 Reducing the Consequences of the Risks

If risk becomes a reality because, for example, the control methods for reducing the likelihood are ineffective, inadequate or unsuitable to prevent the risk, then controls are needed to reduce the consequences or impact of the risk. This includes using detective, responsive, reactive or recovery types of control methods to reduce the severity of the impact or loss caused by the risk.

Information security procedures and processes, information security incident management process, business continuity and disaster recovery plans can all help to reduce the impact of risks.

Lack of management support, commitment and leadership for the ISMS, lack of management reviews, lack of or ineffective performance evaluation exercises all can result in business impacts. The impact on the organization failing to continually improve the suitability, adequacy and effectiveness of the ISMS will mean its ability to protect its sensitive and critical information will decrease and become ineffective. This will result in other impacts that will affect the organization being able to carry out its business properly, productively and profitably.

An information security incident management process detects, investigates, responds, reacts and resolves actual, suspected or potential information security-related events. This process results in actions aiming to avoid or reduce the consequences or impact of the incident or reduce the likelihood of a further reoccurrences of the incident.

Some risks can cause severe disruption to an organization's operations and productivity. Clearly, when such a disruption takes place, the organization wants to recover its business as soon as possible, as downtimes and lack of availability of systems and process can have many impacts such as financial implications in terms of lost revenue; reduced productivity; lower quality levels; reduction in service levels; damage to its business reputation; harms to the trust and confidence its customers have in its business and the potential loss of customers who are not satisfied with loss of services as a result of the disruption.

The risks in some areas can have severe and dire business consequences and impacts. Lack of control regarding compliance with laws and regulations can leave the organization with severe legal repercussions: fines, penalties and legal fees. Protecting staff and customers' personal data according to national legislation around the world is a case in question and suitable measures need to be in place to ensure that compliance with laws and regulations relating to personal data are implemented.

In the area of physical security, computer rooms need to be protected against the risk of unauthorized access. Implementing physical access controls reduces the likelihood of the risk of unauthorized access. However, if these controls are breached, then detective, response or reactive controls are needed to reduce the impact of the risk, for example, alarm systems and CCTV systems.

4.2.3.5 Retaining and Accepting the Risks

This means the organization is able to tolerate and retain the risk within its business risk management strategy by accepting the impact or loss when it occurs. Management must be fully knowledgeable and objective about this course of action. At the end of the day, perhaps there is no other course of appropriate action to take. By default, all risks that are not avoided, transferred, reduced or transferred are retained.

Risk sharing within a group of individuals or companies is also a risk retention option. This involves spreading the risk across the group by transferring the risk liabilities and losses to all those members involved in the group. This is not the transfer by insurance option.

When the risks are small acceptance or retention is a viable and financial strategic option. We need to bear in mind what might be small to one organization may not be small to another in terms of the costs and financial impact they are able to sustain. Acceptance of risk is especially the case where the cost of reducing or transferring the risk could add up in the end to be greater in the long run than the total losses sustained.

4.2.4 Determine the Controls

After the risk treatment options have been selected, then a set of controls need to be determined to implement these options; see Chapter 6 on controls to modify risks. The organization needs to decide which controls it needs to modify the identified risks to a level that is acceptable to the organization. It is important that existing controls in place are taken into consideration. These existing controls may need to be modified if it is decided they are still useful or replaced.

4.2.5 Statement of Applicability

After determining the controls, a gap analysis is then carried out comparing this set of controls with the controls referenced in Annex A. This check aims to make sure that no control areas are overlooked. Note that the process of determining controls does not mean that controls are selected from Annex A. The determination process does not involve Annex A.

The results of the control determination and gap analysis processes are brought together in what is called a Statement of Applicability (SoA). This provides a useful record of the results of risk assessment and risk treatment process. The SoA is also a key piece of documentation for ISMS certification audits (see Chapter 10).

4.2.6 Risk Treatment Plan

A risk treatment plan should be produced that details how the risk treatment options that have been chosen will be implemented by the organization. This plan should provide the reasons why the specific selection of treatment options has been chosen and what the expected benefits are that the organization expects to gain. The plan should detail the priorities for actions to be taken to implement the treatment options, legal and regulatory measures to be taken, the resources required for the implementation and the schedule and action delivery dates for implementation. Those risk owners who are accountable for approving the plan need to be identified as well as those who are responsible for implementing the plan. In addition, performance metrics and measurements need to be determined, and a reporting mechanism needs to be specified.

4.2.7 Risk Owner's Duties

A very important requirement of the standard and the risk assessment process is to identify risk owners, that is, those individuals in the organization that are accountable, and have the authority, for decisions regarding the management of the risks and ultimately regarding managing the risks. So, for example, an individual that is authorized and accountable for personal data of employees and staff would be the obvious person to be the risk owner of such information. Such an individual would be responsible, for example, for signing off the implementation of a set of controls for the protection this personal data and for approving the resulting level of residual risk and for the long-term accountability and risk management of this personal data. The same situation applies to all the other risk owners in the organization.

4.3 Ongoing Reassessment of Risk

4.3.1 Risk Reviews and Reassessments

A risk assessment is a snapshot in time of the risks that the organization faces when the assessment is carried out. The conditions, circumstances and other factors that were applicable to the risks that were identified during this assessment will change at some point in time this is the nature of everything in the world around us (nothing is permanent) constant process of change is a certainty.

Organizations should expect that changes will happen and that this will change their risk profile, and for this reason a reassessment of the risks needs to be undertaken at regular intervals. What is a regular interval? This is up to the organization to determine; however, typically this could be every 6 to 9 months or sooner if there are known or planned changes or there is a major incident or event that has caused the disruptive risks to the organization. Some organizations set their reassessment interval at 3 to 4 months because of the nature of their business environment and the dynamically changing business environment in which they operate.

4.3.2 Risk Monitoring

As conditions and factors that will have an effect on the risk profile of an organization will change, vary and fluctuate, ongoing monitoring is essential to keep up-to-date and for continual improvement of the ISMS to ensure suitable, adequate and effective protection. In Chapter 9, we shall look into more detail the topic of monitoring and measurements. At this point, we shall only provide an overview of the importance of risk monitoring.

4.3.3 Updating the Risk Treatment

After reassessing the risks, there will be a need to reassess the ISMS risk treatment options. This needs to take consideration of whether to take corrective action to improve the existing set of controls, and/or to implement a different set of controls and/or select additional treatment options. This will involve producing an updated risk treatment plan.

CHAPTER

5

Contents

5.1 Management Policy

5.2 Leadership

5.3 Roles and Responsibilities

5.4 Resources

5.5 Training and Awareness

ISMS Leadership and Support

Good actions give strength to ourselves and inspire good actions in others.

—Accredited to Plato

5.1 Management Policy

An important aspect of conformance to the requirements of ISO/IEC 27001 and to achieving a successful ISMS development and implementation and ongoing management is that such aspects should be driven and led from the top, by top management.

Top management should start by defining an appropriate information security policy for the ISMS. The policy should be a clear management statement of its intentions, objectives and goals regarding information security and the protection of its information systems. This policy should reflect top management commitment and support for the ISMS to satisfy the requirements and address the issues in ISO/IEC 27001 Clauses 4.1 and 4.2. It should be a directive from top management that typically should address:

- The scope of information security, its importance to the business and making it clear what the business information security objectives are (e.g., regarding the protection of the confidentiality, integrity and availability of its information);
- The need for staff awareness: make staff aware of their duties and responsibilities of the risks (e.g., their responsibility to handle and process company commercial-sensitive information in way that protects it from compromise);
- What is acceptable and not acceptable with regard to behaviour and the use of its resources (e.g., acceptable use of the company email system);
- Its obligations to carry out its business in compliance with the laws and regulations, contractual obligations, best practices and standards with which staff also need to comply (e.g., compliance with laws on copyright, data privacy or protection and computer use, misuse or abuse);
- Reference to any other documents that staff need to be aware of and comply with, for example, more detailed security policies and procedures as well as any other relevant proceedings not those directly related to security; this could be industry-specific policies such as those businesses that need to deal with environmental issues, aspects of healthcare, production of pharmaceutical products or food safety.

This management information security policy should be written in a way that the style and content are independent of any particular skill, process or technical knowledge. For example, the content should be understandable by someone who is not an IT specialist, someone who is not trained in company finances or legal affairs or someone who does not have human resources skills. In other words, it should state information security objectives that are generally understood by all staff, not just someone with a highly technical background or certain professional qualifications or skills.

5.1.1 Approval, Communication and Awareness

This management policy needs to be approved and signed off by top management (or someone of similar management authority and accountable status) because the aim is to indicate management commitment and support.

The policy needs to be communicated to all management and staff across the organization and interested parties where necessary and appropriate. This could be in a paper form or by electronic means or both. Some organizations display their policies on the walls of offices, computer rooms and other areas to ensure that they are continually accessible and visible. Other organizations resort to using information and communication technology (ICT) to distribute and have available their policies and procedures via their internal network. Others may choose to distribute it in paper only for staff to keep at their own place of work. Whatever the method used, the policy should not be hidden away and forgotten. Staff need to read, understand and refresh their memories regularly of its contents and what it says about their specific information security responsibilities and duties.

5.1.2 Policy Review

This policy needs to be reviewed and updated as and when necessary to take account of the changing nature of the information security risk environment and evolving organizational developments and changes. There needs to be a review process for maintaining this policy, as is the case with all policies and procedures, as part of the ISMS continual improvement process.

5.1.3 Management Policy Sets the Scene

This management policy is a high-level policy that "sets the scene" and typically there will be more detailed policies (covered in Chapter 7), which will give more specific rules and instructions on the implementation of information security protection. For example, policies on access control cover the rules of access to different organizational resources and facilities such as email servers, databases, network services, applications and physical access to buildings, offices, rooms and storage equipment.

5.2 Leadership

Demonstrating leadership regarding the ISMS is an essential aspect of ISO/IEC 27001 as a management standard. It is of the upmost importance that top management provides the appropriate level of leadership in terms of direction, authority, policy, governance and organizational control. Good leadership defines business purpose and context for information security; creates the mission statement and sets the strategy; provides staff focus on what is important with regard to the information security for the business and what the priorities are; motivates and inspires confidence and trust in the workforce that it is committed to protecting the business; and nurtures security culture and security skills.

Good ISMS leadership is needed to build a team that will successfully take forward the implementation of the ISMS, which will empower and motivate staff to be proactive followers and supporters in helping to protect the organization. A good ISMS leader will be passionate about being successful in managing the information security risks the organization faces. ISMS leadership should strive to inspire others to see information security as a business enabler, with the vision of turning information security risks into a business opportunity.

Leadership is different to management; the former motivates and inspires, creates the vision and points people in the right direction, and the latter administers, controls and follows the vision and organizes people. Both ISMS leadership and ISMS management together achieve an effective, robust, resilient ISMS. Leadership will be the champion of the ISMS and management will control and manage the ISMS.

5.3 Roles and Responsibilities

Day-to-day working and operational activities functioning effectively, and the proper management of staff, at all levels throughout the organization, can contribute to a good information security business environment. In part, this requires a good information security culture within the organization to be in place, with appropriate awareness and understanding of the problems of information security risks and clear lines of responsibility and accountability.

It is essential that roles and responsibilities for protecting specific types of information or information systems or for carrying out specific information security-related processes are clearly defined and allocated. For example:

- Owners of an information system should be given the information security responsibility and accountability for that system (these owners may delegate that day-to-day implementation of security to another individual or to a service provider, but they remain ultimately accountable for the protection of the system and the management of the information security risks);
- Data owners, data custodians;
- PII controllers and processers;
- Business owner: a specific department or group that ensures implementation of policy and procedures, defines information usage and classification for information in their custody, allocates information custodians and defines access roles and privileges, staff training and awareness and protection of personal data under their control;

- Chief information security officer (CISO), information security analysist, information security team;
- Information security incident response team;
- Business continuity manager;
- Internal auditors;
- Human resource manager;
- IT services manager (IT service management, IT disaster recovery, involvement in incident management);
- IT and network administrators or managers (network management, secure network technologies, involvement in incident management);
- Users of information systems.

In addition, all staff will have general responsibility of information security related to their day-to-day work. For example, reporting of unusual or suspicious behaviour, either related to their use of IT or network services or related to other staff, contractors or visitors. Also, all individual staff needs to be aware of their responsibility of protecting the information that they handle, share and process, using their access rights and privileges correctly, keeping safe their access credentials such as their passwords and other types of access codes and ensuring that they are using organizational resources in accordance with the acceptable usage policy, for example, rules for using email for sending file attachments.

5.4 Resources

As with any successful business venture, it is important to have the right type of resource for the jobs that need to be done. Having staff with the right competence to do a job properly, efficiently and effectively is key to the overall success of the business. If it is a technical job, then the staff involved need to have the right level of knowledge and skill to handle technical requirements of the job at hand, to resolve technical problems and to be able to use techniques, methods, equipment and procedures relevant to the technical area in question. If it is a customer services job, then it requires the staff involved to have the relevant skills needed to deal with customers; for example, they are able to listen and respond effectively to customer's questions and queries, and they are able to satisfactorily resolve customer queries and problems, follow up on feedback from customers and generally be able to meet the expectations of the organization's customers.

Management needs to ensure that, for specific information security tasks, it has the right people, with the right skills and knowledge and experience. This can mean recruiting people who have the right existing skills and experience or recruiting people and providing a training programme for them to develop the right skills and experience. In addition, all staff working in the field of information security, whether those with experience or those in training, need to keep up-to-date as the issues and risks in information security continually evolve, as does technology and business practices.

Every organization strives to have human resources with certain core competence and with regard to organizations seeking staff with information security skills, the market is expanding, becoming more buoyant and highly competitive. For many years, organizations have recruited information security personnel based on hands-on experience, practical knowledge and appropriate references. Even though this is still the general basis of recruitment, more organizations have started to request that applicants have market-proven professional qualifications, personal certifications and, in some cases, university qualification or a combination of both.

The current education and training market provides a range of certified qualifications, for example, in areas such as:

- Information security auditing;
- Information security management;
- Risk management;
- Information assurance;
- Governance;
- IT security;
- Physical security.

Many universities around the world are now offering information security and IT security modules as part of BSc degrees; some are offering MSc degrees in information security, and some run PhD programmes in information security research.

The market for information security education and training is growing rapidly, which reflects the increased awareness of the risks and importance of protecting information. This trend means that, in the near future, organizations will have access to a greater number of human resources with professional qualifications to complement the practical on-the-job experience and skills needed from information security practitioners.

5.5 Training and Awareness

The organization needs to ensure that top management, management and staff are aware of information security risks and have sufficient understanding to support the organization's information security policy and to undertake their normal work functions and tasks. There are at least four types of training: basic awareness (applicable to all); management training; compliance training; and technical training (applicable to those needing technical skills and knowledge of information security).

All staff should be trained in the use of the ISMS information security policy and related processes and procedures, about keeping information safe and secure, a basic understanding of compliance with laws and regulations, a basic knowledge of security measures and controls for modifying risks applicable to their job function and in the correct use of IT (for example, log in procedures, keeping passwords safe, use of one-time passwords and appropriate use of IT). Beyond this basic level, some staff will need more technical training.

5.5.1 When Should Training Take Place?

Induction training for new staff should take place on joining the organization. This should cover the company's information security policies, procedures and routine practices, to whom to contact for help, support for information security matters, to whom to report security problems, initial familiarization with the common types of risk malware, hacking and protection of commercially sensitive information and the protection of personal data, fraud and use of email.

On-the-job training provides specifically tailored instructions on information security suited to the individuals job function. There are upgrades in training for those whose job function changes, and they might need a more technical level of training. Refresher training keeps staff up-to-date with new developments and to provide an organization-wide reminder or more immediate remedial training as the result of a security incident or an emerging risk.

5.5.2 Training Methods

An organization can deploy a variety of methods to deliver effective information security awareness and training to its staff. The methods that an organization selects will depend on the business culture and its operational needs. Therefore, an information security awareness and training programme should be tailored to the specifics of the organization. You should alternate between different methods, perhaps introducing an element of fun into the mix, but always a degree of interactivity.

- Classroom-based training can be highly interactive; such training can vary from a half-day to 1-day introductory course or beginners' training course through various intermediate or advanced 3-to-5-day training courses covering a range of specific topics.
- Computer-based or online web-based training and awareness are a good method for reinforcing information security principles and specific topics. Such training can be delivered as a set of modules, interactively or noninteractively, and can be accessible to staff at a time and place convenient to the individual.
- Seminars, workshops, roundtable discussions and presentations are especially well-suited for introducing new subject matter and for organizations with multiple sites.
- Videos are also an effective way to provide training on various topics.
- Posters provide visual reinforcement of information security principles and specific topics.
- On-the-job or desktop training can be helpful.
- Internal emails can be used to remind, reinforce and provide updates on organizational policy and procedures.

5.5.3 ISMS-Related Topics

Implementing the requirements of ISO/IEC 27001 covers many different tasks, activities and processes that need to be carried out and these are associated with several specific topics that information security professionals, practitioners and other staff will need to have knowledge of and develop experience in, depending on their particular job function. For example, the requirement for an internal ISMS auditor will be knowledge of the auditing process and methods, whereas an ISMS risk manager would need knowledge and skills of the principles of risk management. These include:

- Principles of risk management (e.g., as found in ISO/IEC 27005, NIST SP 800-30, NIST SP 800-37, NIST SP 800-39):
    - Risk assessment;
    - Risk treatment.
- Performance evaluations:

    - Measuring and monitoring methods and techniques (e.g., as found in ISO/IEC 27004);
    - ISMS auditing process and methods (e.g., as found in ISO 19011 and ISO/IEC 27007).
- Legislation and regulations related to information security and privacy:
    - National and regional laws and directives (e.g., EU GDPR);
    - Standards covering privacy (e.g., ISO/IEC 29100, 29134, 29190).
- Information security controls: for example, 27002 (generic), 27011 (telecoms), 27017 (cloud security), 27018 (cloud privacy), 27019 (energy sector security), 27799 (health-informatics security), NIST FIPS 200, NIST SP 800, COBIT 2019 or later versions, SANS CIS 18;
- Certification and auditing:
    - Certification (e.g., as found in ISO/IEC 27006, ISO 17021-1);
    - Auditing (e.g., as found in ISO/IEC 27007, ISO 19011).
- Other specific security-related processes:
    - Incident handling (e.g., as found in ISO/IEC 27035, ISO/IEC 27041, ISO/IEC 27043, NIST SP 800-61);
    - Business continuity (e.g., as found in ISO 22301, ISO/IEC 27031);
    - Operational resilience (e.g., as found in ISO 22316).
- Specific IT security controls and mechanisms:
    - Network security (e.g., as found in ISO/IEC 27033),
    - Applications security (e.g., as found in ISO/IEC 27034),
    - Malware protection (e.g., as found in NIST SP 800-83),
    - Firewalls, IDS, IPS (e.g., ISO/IEC 27039, ISO/IEC 27002);
    - Access control methods (e.g., ISO/IEC 27002, ISO/IEC 27040).

CHAPTER

6

Contents

6.1 Determining the Controls

6.2 System of Controls

6.3 Control Through Policies and Procedures

6.4 Examples of Control Sketches

6.5 Sector and Application Specific Controls

6.6 ISO/IEC 27001 Annex A

Measures to Modify the Risks

An ounce of prevention is worth a pound of cure.

—Benjamin Franklin

6.1 Determining the Controls

6.1.1 Control Framework

The outcome of the risk assessment and treatment process should be a set of controls for managing the risks. The framework for achieving this is described in Clause 6 of ISO/IEC 27001. As mentioned in Chapter 4, risk treatment involves determining an option or options for risk treatment, determining a set of controls, do a comparison with Annex A, producing a statement of applicability, producing a risk treatment plan, approval and sign-off of plan and residual risks. The process also includes a checking stage to make sure no important areas of control have not been missed during the determination stage.

In order to claim valid conformance to ISO/IEC 27001, these risk management requirements must be fulfilled. Hence, Subclause 6.1.3(b) states that the organization must determine which controls are suitable and appropriate to manage its assessed risks. Therefore, this set of controls will be individual to the organization and in many respects likely different to another organization in the details of the controls.

If there are controls in Annex A that are not included in the set of determined controls, then justification as why they have not been identified for implementation needs to be given.

6.1.2 Process of Determining a Control Set

Controls are determined to implement the risk treatment option selected based on their capability of modifying risks. Also, a control or a set of

Important Note

1. In the 2005 edition, the controls were selected from Annex A of the standard; however, this was changed in the 2013 and 2022 editions. This points to a major difference between the first edition (2005) of ISO/IEC 27001 and the second and third editions (2023 and 2022) in the risk treatment process. Annex A is still retained in the standard, but its function and purpose have completely changed and is strictly limited to a checking function. The purpose of Annex A in the 2013 and 2022 editions of the standard is as a benchmark to check that the controls that have been determined, by design or identified from other sources, do cover, as a minimum, the set of reference controls addressed in Annex A. Controls are determined (Clause 6.1.3) by organization design or from other sources (e.g., codes of practice, international and industry standards, sector specific standards). This change in emphasis from the selection of Annex A controls to determining controls from any source provides greater flexibility and choice for organizations to integrate with other internal control and standards and harmonize with business sectors with which the organization works. This presents less restrictions and limitations on the organization and more freedom without being tied to Annex A.

2. Users of ISO/IEC 27001 wishing to claim conformance need to use the current edition, which, at the time of this writing, was the 2022 edition. The 2005 and 2013 editions are no longer valid standards and no longer supported by ISO.

controls in combination may be used to prevent, detect, reduce or correct the undesirable effects of an information security incident, breach or compromise. Controls that are used to prevent an information security incident include those controls that reduce the likelihood of the incident occurring in the first place. For example, top management providing sufficient resources to ensure an effective ISMS is in place is a preventive measure. Another example is a well-written and properly deployed disaster recovery plan and procedure can reduce the consequences or impact on the organization in the event of a disaster or business disruption. Also, the use of firewall technology is a further example.

Some controls are able to detect indications that an information security incident or attack is about to happen or occur. If such detected indications are effectively acted upon in a timely manner, they can reduce the potential impact of the attack and, in some cases, stop the attack from occurring in the first place in essence reducing its likelihood. For example, having malware detection software implemented to prevent infection of IT systems and subsequent loss, damage or destruction of information.

If the knowledge gained by an employee from an information security awareness and training programme is properly put into practice by the employee in their everyday jobs, then this contributes to reducing the likelihood and impact of an attack. For example, if employees know when, how and why they need to report about potential events that might be a precursor or an indicator of an information security attack or incident, then this helps in the analysis and response, which subsequently can help in reducing the potential impact of the attack. Equally, if employees know what to do to protect the sensitive information, they are processing by following a company procedure that gives instructions of how to protect such information, then they are contributing to reducing the likelihood that sensitive information will be leaked or stolen.

Therefore, in determining the controls needed to modify the risks identified with regard to the treatment options, reduce the consequences, and/or the likelihood of the risk; take into account the function of the control to the extent that the control:

- Reduces the likelihood of a risk-related ISMS attack, incident or event occurring (for example, which can mean reducing or removing the weaknesses or vulnerabilities in the business systems or processes, improving the effectiveness of a control to prevent an information security incident being realized, to implement a control to deter an information security incident happening).
- Reduces the consequences or impacts of an attack, incident or event (for example, which can mean implementing controls that will re-

spond, react to an information security incident that is happening or is expected to happen to limit or reduce the consequences of the incident to implement a control to correct the effects and impacts of an information security incident happening).

- Detects and responds to precursors and indicators of an attack, incident or event.
- Implements preventative actions against the future occurrence of an attack, incident or event.
- Implements corrective actions to remove the cause of a past attack, incident or event.

6.1.3 Control Sources

ISO/IEC 27001 6.1.3(b) states controls can be determined from any source. This might be controls specified in a National Institute of Standards and Technology (NIST) standard or the Information System and Audit Control Association (ISACA) Control Objectives for Information Technologies (COBIT) standard. There are many control standards that have been defined and specified by various organizations, business trade groups and association and standards bodies.

This means the controls do not need to be from ISO standards; they can be from other sources. They can be designed and developed in-house by the organization, adopted from examples of existing controls that have been successfully used by other organizations and industries. The important thing to have in mind is that the controls are appropriate and suitable to control and mitigate the risks your organization faces.

An example set of published sources are:

- ISO and IEC and ITU-T standards;
- NIST Computer Security Resource Center, https://csrc.nist.gov:
  - SP 800-53—Security and Privacy Controls for Information Systems and Organizations;
  - NIST SP 800-171;
  - Cybersecurity Framework (CSF);
  - NIST SP 1800 series.
- ISACA;
- COBIT: https://www.isaca.org/resources/cobit;

- CIS (Center for Internet Security): Critical Security Controls, https://www.cisecurity.org/controls;
- National Cyber Security Centre (NCSC): https://www.ncsc.gov.uk:
    - Cyber Assessment Framework (CAF);
    - Cyber Essentials;
    - Cloud Security Principles.
- General Data Protection Regulation (GDPR): Framework of security requirements that global organizations must implement to protect the security and privacy of EU citizens' personal information: https://gdpr-info.eu;
- Cloud Security Alliance (CSA): https://cloudsecurityalliance.org; Cloud Control Matrix (CCM): https://cloudsecurityalliance.org/research/cloud-controls-matrix;
- International Security Forum (ISF): Standard of Good Practice for information security, https://www.securityforum.org/solutions-and-insights/standard-of-good-practice-for-information-security/.

6.1.4 Existing Control Sets

In determining which controls are needed to modify the risks identified in the risk assessment process, it is also important to take into account any existing controls the organization may have implemented. Any system of controls needs to work together and so it is necessary to understand what controls already exist and what role and effect they have in modifying the risk. It is also important to check whether any new controls that are planned for implementation are likely to conflict with the existing set, whether there is redundancy or how effectively will they work together and interwork.

It is likely that an organization will already have some controls in place prior to implementing ISO/IEC 27001. For example, an organization probably has at least some access control methods in place, but in doing a risk assessment, it is found that these methods are not sufficient so they need to be either enhanced or replaced. Another example might be that the organization has controls in place relating to employee training, procedures for handling sensitive information or for secure remote working. All of these need to be considered when doing a risk assessment and in determining a set of controls.

6.2 System of Controls

6.2.1 Control Framework

Controls are determined as part of the risk treatment process on the basis that they can modify the risks identified in the risk assessment process. Controls can be of many forms such as policies and procedures, processes, methods or techniques. It is important to bear in mind that ISO/IEC 27001 is a management system standard and the object of protection is information, so the controls are oriented toward management and organizational requirements for information security and not IT or technology-oriented. However, as IT is today a predominant tool of modern business management, operational use will need to be considered in the protection of information.

Figure 6.1 shows a typical grouping of controls, which can be related to management systems.

6.2.2 Requirements and System of Controls

The requirements of ISO/IEC 27001 are either management or organizational in nature, with the prime objective of protecting information assets. As such, controls should be driven by these requirements with the end goal of protecting information and all forms of information asset. For example, the provision of resources by the organization (Clause 7 of ISO/IEC 27001) for the ISMS, which involves personnel and communications. Another example is top management establishing an information security policy (Clause 5.2 of ISO/IEC 27001), which includes addressing information security objectives and requirements at relevant functions and levels. In addition, there are the requirements for the organization to manage its

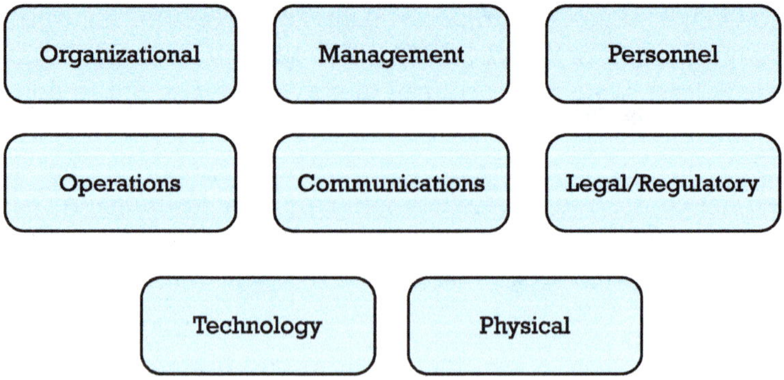

Figure 6.1 Control framework.

risks (Clause 6 of ISO/IEC 27001). Hence, there is a wide scope associated with the control framework indicated in Figure 6.1.

6.2.3 Examples of Groups of Control

6.2.3.1 Organizational Policy, Governance, Risk and Compliance

ISO/IEC 27001 is an organizational standard to assist top management to manage and govern its business to protect it against information security risks and helps its compliance with laws and regulations. Controls at the organizational level are essential for the overall well-being of the company and for its protection against information security risks. Leadership requirements (as specified in Clause 5 of ISO/IEC 27001) are an essential area of protection that needs appropriate controls to be in place, which include policy, roles and responsibilities for information security, management responsibilities and committing adequate resources, taking into account the context of the organization as presented in Clause 4 of ISO/IEC 27001.

ISO/IEC 27001 specifies only one information security policy; this is in Clause 5.2. This is an all-embracing policy setting security objectives or providing the basis for defining these objectives, taking into account all security requirements and committing to continual improvement of the ISMS.

Some organizations have additional policies covering more detailed rules and requirements at the technical level related to specific topics (for example, access control policy), but these additional policies will need to conform with the organization policy.

6.2.3.2 Organizational and Management Procedures and Processes

The organization needs procedures in place to inform employees of the correct and proper way to do something to ensure information security is maintained, for example, an employee procedure for processing customer data to ensure protection of the data. Other examples might include rules and procedures for the acceptable use of technology and resources to ensure, for example, transferring company information (including personal data) through email or other messaging applications. Organizations need to deploy arrangements for engaging with government and regulatory authorities and special interest parties, business associations dealing with information security. Many organizations deploy a process for threat intelligence, to gather and assess the threat environment to enable appropriate mitigation to be implemented.

Several procedures need to be considered to manage the ISMS, for example, assess management, information classification and labelling management, for the segregation of duties, security protection in project management, access and identity management, access rights, credentials and information for authentication.

The organization also needs processes in place to ensure the continual improvement of the ISMS, for monitoring the performance of the ISMS and the ongoing assessment of risk. This should include management reviews, internal ISMS audits and security assessments (if applicable). They also need an asset management process to be implemented.

In the same way that an organization will have an overarching organization policy with possibly several more technically detailed supporting policies, the same applies to procedures: for example, an organizational procedure for processing customer data might be supported by more detailed procedures operating at the technology level.

Effective and timely communications for sharing information and updates with employees and business partners are also essential.

Procedures for outsourcing and supply chain management arrangements and supplier services need to implemented. This includes information security included in supplier agreements and contracts, incident reporting, monitoring and reviewing practices used by suppliers with regard to information security and managing information security associated with the technology used in the supply chain.

6.2.3.3 Personnel Controls

People in an organization are key to ISMS protection, but, at the same time, they can also be the greatest weakness to this protection; as the well-known, often-quoted phrase says, "people are the weakest link." Insider threats, social engineering and other situations involving people support this weakest link quote. Employees should be considered by the organization as their greatest asset rather than a weakness in which case the organization needs to invest in well-trained and aware employees.

The unaware or uninformed employee can be the weakness that triggers a security incident. One of the most important aspects of having an effective ISMS is for the organization to have a competent workforce, and this requires the workforce to be adequately educated, trained and aware. This means trained and aware in the policies, procedures and processes related to the ISMS and their job function, also the threats and risks related to the ISMS in their day-to-day work.

There should be measures and employment practices relating to security regarding the recruitment: before, during, internal change of job function and terminating employment. Other measures include disciplinary process, user access registration and deregistration process and employee incident reporting.

6.2.3.4 Operations and Business Communications

Smooth and uninterrupted operations are what most organizations want to achieve. Incidents can cripple an organization or severely disrupt its operations and cause damage to its reputation and its data and result in financial losses or penalties and fines. Incident management, business continuity and disaster recovery are all essential aspects of the operations of an organization to avoid untold problems and loss of business. Operations need controls in place covering day-to-day activities to ensure the continued business. This includes regular monitoring of the risks and taking the necessary action to mitigate them, detection controls for incident and events, regular backups of data, maintenance of systems and resources, updating software, configuration management, change controls covering the updating of resources and operational practices and procedures.

Communications security is another area with which organizations need to be concerned. Controls for protecting information being communicated when using the internet and messaging platforms, and using social media for business purposes, should also be in place.

6.2.3.5 Legal, Regulatory and Contractual Controls

Over the last 3 to 4 decades, there have been more laws and regulations published that are relevant to information security. This includes the growth in laws to address the issue of privacy, that is, to protect personal data or PII. Protection of personal data also relates to the safety of users online and online identities and the related security problems involving online fraud, scams, cloning, harassment and abuse.

More areas of business and society are being influenced and dominated by the use of technology, which bring about challenges and issues of abuse and exploitation. Some of the areas are now the focus of lawmakers; for example, social media, online banking and other internet and Wi-Fi-related applications are being exploited in a troublesome and negative way. Cyber legislation is now the focus of many countries, for example, the European Union, United Kingdom, United States, China and Japan.

6.2.3.6 Technical Controls

Information being processed, stored, archived, communicated and shared needs protecting; this is a prime objective of information security and ISO/IEC 27001. Technical controls for ISMS should be related to the protection of information assets of the organization not about protecting its technology. What this means in today's technology-driven world is that controls need to be in place to protect information being processed, stored, archived, communicated and shared using technology. Access to information in a digital form is a good example where controls are needed to prevent unauthorized access to information through the use and application of technology, in a similar way to having control of access to information in a physical form. The difference being that information in digital form has more variants than its physical counterpart. Reiterating the fact that ISO/IEC 27001 is not about IT security but information security reminds us that the technical controls needed for the ISMS are about managing the use of technology rather than, for example, the details of the security of technical specifications. Indirectly, if the technology is not safe and secure and is vulnerable, then this can put information at risk. The ISMS is concerned, for example, with using technology for access management, identity management and authentication techniques and for remote working and giving access to mobile services, use of antivirus software, backups, managing network security and other control areas related to the organization's use and management of technology.

6.2.3.7 Physical

The organization needs to physically protect its information assets and resources. This includes controls protecting physical resources and facilities that contain and stores information assets: buildings offices, rooms, equipment and storage media. Physical entry controls and physical perimeter or boundary controls around buildings or offices. Controls for dealing with visitors that are given access to buildings or offices. Storage equipment and media need physical access controls. Communications equipment and cabling protection needs to be considered. Server rooms need protection. Security controls for the disposal of storage media and computer equipment need consideration, as do equipment maintenance and procedures and controls for working in designated secure areas and areas with restricted access, for example, server rooms or rooms dedicated to a group of employees working with confidential information.

6.3 Control Through Policies and Procedures

6.3.1 General

This section presents a brief description of policies and procedures that provide control of the risks to information security. This is not a definitive account of such controls, and more detailed commentaries can be found in other sources. These controls are introductory in nature and, in Section 6.4, examples are provided of control sketches.

6.3.2 Policies and Procedures

Information security policies can either be a high level of detail or a more in-depth level of detail. The requirements of ISO/IEC 27001 only specify (in Clause 5.2) one type of policy, which is a top management and high-level policy. This policy is a central part of the ISMS protection strategy and mission statement and is a mandatory requirement.

An organization may, based on the results of its risk assessment and treatment, also decide it needs to have policies and procedures at different levels of details targeted at specific topics such as mobile device policy, access control procedures, clear desk policy, backup procedures, procedures for using social media, remote working procedures and policy of the use of cryptographic controls. These policies are not requirements of ISO/IEC 27001 but may be deemed necessary as a result of the risk assessment and as a control determined suitable to mitigate the identified risks.

The top management information security policy (ISO/IEC 27001, Clause 5.2) is intended to be a high-level directive and mission statement appropriate to the whole organization and its business and applicable to all the requirements of ISO/IEC 27001. Such a policy includes the information security objectives, a commitment to satisfy the requirements of interested parties (ISO/IEC 27001 Clauses 4.1 and 4.2) and an undertaking to continual improvement of the ISMS (ISO/IEC 27001 Clause 10).

This information security policy should be written in a way that the style and content are independent of any particular skill, specialist or technical knowledge; it should be communicated to all in the organization and to other interested parties as and where appropriate. The content should be understandable by someone who is not a specialist, not trained in company finances and not having human resources skills. In other words, it should state information security objectives that are generally understood by all staff, not just someone with a highly technical background or certain professional qualifications or skills.

The length of the documented policy (number of pages) is not specified in the standard. Some organizations produce such policies of between 1 and 3 pages; others may decide to produce a policy which is greater in length and volume of text. It is up to the organization to decide the length, volume and detail of text in its policy commensurate with its business objectives, strategy and culture. What is more important is that it contains enough information to convey in a clear way the objectives and ideas referred to above.

The broader aspects of information security policy at all levels of detail should be to

- Provide a clear definition of the scope of information security and its importance to the business and make it clear what the business information security objectives are (e.g., regarding the confidentiality, integrity and availability of its information).
- Make staff aware of their duties and responsibilities with regard to protecting the information over which the organization has control (e.g., their responsibility to handle and process company-restricted and confidential information in way that protects it from compromise).
- Set out what is acceptable and not acceptable with regard to behaviour and use of its resources (e.g., acceptable use of the company email system).
- Make clear its obligations to carry out its business in compliance with the laws and regulations, contractual obligations, best practices and standards with which staff also need to comply (e.g., compliance with laws on copyright, data privacy or protection and computer use, misuse or abuse).
- It should give reference to any other policy documents of which staff need to be aware and comply with, for example, more detailed security policies and procedures as well as any other relevant proceedings, not those directly related to security; this could be industry-specific policies such as those businesses that need to deal with environmental issues, aspects of healthcare, production of pharmaceutical products, or food safety.

6.3.3 Approval, Communications and Awareness

Information security policy needs to be approved and signed by someone of management standing, who has the authority and is of accountable status. This achieves one of the aims of information security policy, that is, to indicate management commitment and support.

Information security policy should be communicated to all staff. Some organizations display their policies on the walls of offices, computer rooms and other areas to ensure they are continually accessible and visible. Many organizations today resort to using electronic means to distribute and have available their policies and procedures via their internal network. Others choose to distribute in paper only for staff to keep at their own place of work. Whatever the method used, the policy should not be hidden away and forgotten.

Staff not only need to be aware of the policy, but also understand its content and what it says about their specific information security responsibilities and duties. Staff need to read, understand and refresh their understanding on a regular basis the contents of policies and procedures.

6.3.4 Review

Like most things in business, there will be changes such as how the organization operates and does its business, the business environment, risk profile and other contexts. Therefore, such changes should be reflected in the information security policy. A review should be made of the policy whenever it is most appropriate, possibly after identifying changes to the business, after an incident or as part of a regular management review on its ISMS (ISO/IEC 27001, Clause 9.3).

6.4 Examples of Control Sketches

6.4.1 Overview

This clause provides some examples of other controls. These examples provide details of what might typically be included in procedural controls. The examples are not the finished product of such controls; they are merely sketches and examples and should be read and considered as such. The finished product would need to be customized by the organization itself and its requirements and objectives. In some cases, for example, the acceptable use policy, the organization would need to have the wording of such policies and procedures be produced by management and checked by legal experts and human resources, as the content and wording of such policies will depend upon the jurisdiction of the country in which they are used and applied. Additional information on operational controls, such as backup process, information security incident management procedures and business continuity plans, will be given in Chapter 8.

6.4.2 Information-Handling Procedures

6.4.2.1 Basic Idea

The aim of these procedures for information security is to ensure that all staff know what they should do to handle information in way that protects its confidentiality, integrity and availability: be it the processing, storage and archiving, distribution, copying or disposal of information.

6.4.2.2 Information Management

- *Ownership, accountability and third-party custodianship:* Information that is generated and/or created using the organization's resources (staff, processes and systems and by other means) to carry out its business is generally deemed to be its property and the organization becomes the owner. The organization might also process information from third parties (e.g., clients, customers, business partners).

  Often, responsibility for the day-to-day management of the information is delegated down to particular groups, departments, functional units and/or individuals within the organization. Custodians within these groups are then given individual operational responsibilities of being the owner/custodian, which means they need to manage these assets and to ensure they afforded an appropriate level of protection. In the area of data protection (personal data or PII), legislation defines the roles of data controller and data user; both these roles are associated with different information responsibilities. It might also be the case that information is managed and processed by an outsourcing contract to a third party. In all cases of delegation, custodianship and outsourcing the ultimate accountability of these assets rest with the data owners, top management, directors and executives, and it is their responsibility to ensure their safe-keeping.

- *Use and misuse of information:* The ISMS involves an ongoing process of practicing the right things to do for the business with due care and diligence to protect information from a range of threats and risks. This means that staff must use the organization's information and resources in an acceptable and proper way to avoid security comprises and/or legal complications and actions; see Clause 6.4.2, "Acceptable Use Policy."

- *Information classification:* Information should be graded appropriately according to its level of sensitivity and/or criticality. This enables an organization to be able to deploy the right amount of protection to the information commensurate with these sensitivity levels. There are many names and levels of grading that can be assigned to information, for example, in confidence, in strictest confidence, commercially confidential, financial in confidence, personnel-in-confidence and restricted/secret.

6.4.2.3 Information Gathering, Creation and Processing

- *Gathering and collecting:* Gathering and collecting information are normal functions of most organizations through their business activities. This may be through market research, research on customer activities (types of spending, seasonal fluctuations), customer surveys and feedback and other forms of research. There are some legal, contractual and security implications about collecting information which organizations should be aware of and pay attention to. The most prominent one is personal data: what can be collected, what can be done with this data and what controls are needed to protect it against abuse or misuse. Indicative of another area of attention is that of intellectual propriety and copyright related to both electronic and nonelectronic information and digital rights, digital rights management (DRM) and the rights of individuals.

- *Processing efficacy, effectiveness, efficiency and correctness:* The activities of processing information can mean many things, but all have information security requirements to ensure the confidentiality, integrity and availability of information are preserved, which means doing the "right" things at the "right" time (efficacy of processing), getting these things done (efficiency of processing) and doing these things "right" (effectiveness of processing).

  Information that is processed by organizations is vital to their business, whether it be processing customer orders, accounts and statements, online transactions buying and selling products, formal company accounts and tax records or other business processes; it is important to get it right to avoid losing money, time, productivity, credibility and possibly customers through incorrect data entry and processing of this data through subsequent results with mistakes and errors. Some of these problems are caused by human errors, lack of training in using processing applications, software flaws, system malfunctions and failures, data corruption and data loss or unauthorized modification.

  It is estimated that a high proportion of business losses and security compromises result from the people who handle information. The correct processing of information is a major area of security problems resulting in customer dissatisfaction, business losses, inefficient and ineffective business operations and legal action for failure to honour contracts or to be compliant with legislation.

- *Off-site processing:* Some information and data might be processed off-site such as a member of staff using a laptop to process information whilst out on business at a client's office, at home, or in a public place. A managed services company, under an outsourcing contract, might process the information. Whatever the business case, the protection of this information during processing should be afforded at least the same level of protection, as it would receive on-site.

6.4.2.4 Information Storage

- *Electronic platforms and equipment:* With today's technology, organizations and users have many ways of storing information:

  (a) Network servers;

  (b) Cloud platforms;

  (c) Computer hard drives (internal and external);

  (d) Optical disks (e.g., CDs and DVDs);

  (e) Flash memory devices (e.g., USB sticks, SD cards, memory cards), CompactFlash cards;

  (f) Tablets, iPads, iPods and MP3 players and other similar technologies.

  Whatever means and methods of storage are used, security should be addressed: ensuring that confidential information is protected to prevent unauthorized disclosure, the integrity of the information to ensure the information is not changed, modified or flawed whilst in storage and that the information is available to all those who need it. A clear policy should define what protection measures need to be in place to protect its confidentiality, integrity and availability. The more sensitive and/or critical information should be stored using memory segmentation or paging, cryptographic technology and/or a strong access control system.

- *Paper-based:* Information in paper-based form requires just as much attention as information in electronic form. This should include:

  (a) Physically protecting sensitive and/or critical information by locking papers, documents and reports in lockable filing cabinets, desk drawers or safes, which reduces the risk to an appropriate level, as well as the normal physical security to the building, offices and rooms;

  (b) Physical separation of papers with different information classifications;

  (c) Establishing a clear desk policy to ensure papers and documents are put away when unattended;

(d) Protecting papers and documents from physical or environmental threats and damage such as fire and water;

(e) Giving the same level of physical protection to papers and documents kept off-site.

- *On-site and/or off-site:* The decision of whether information should be on-site or off-site depends on the circumstances of the organization: whether off-site storage is a better option and whether it affords a better level of risk management taking into account the cost of such storage and the availability and convenience of accessibility to such storage. Off-site storage should always have the same level of protection as that given to on-site storage.

Information might be looked after by a managed services company that stores and processes information on behalf of the organization. Access to this information is facilitated by the use of networking.

6.4.2.5 Information Sharing, Exchange and Distribution

- *Electronic communications and messaging:* With today's technology, organizations and users have many ways of distributing, sharing and communicating information. Users can push information out (such as web-casting, broadcasting or server-push) to recipients or have available on a network to enable receipts to pull the information down from a website or server. Information can be shared between devices through different messaging platforms, using Wi-Fi, Bluetooth or social media applications.

  Whatever means and methods are used, information security should be a foremost issue to be addressed that:

(a) Distribution of sensitive and/or critical information is kept to a minimum and the information should be clearly marked to indicate its classification.

(b) The organization ensures confidential information is protected to prevent unauthorized disclosure or interception.

(c) The organization ensures that the integrity of the information in that it does not get changed, modified or flawed whilst on route.

(d) The information is available to all those who need it.

(e) A clear policy needs to be established on what types of information can be sent electronically (e.g., via email, email attachments) and what protection measures need to be in place to protect the more sensitive and/or critical information.

(f) If confidential, it can be sent by those with the right level of authorization and then cryptographic technology should be used.

(g) The use of anonymous re-mailers and other technologies should be forbidden.

- *Voice communications:* Communicating via phones and face-to-face needs appropriate attention to what information is being shared, both business and personal, and whether the communication is in a public or private place. Today, videoconferencing or online conferencing platforms are common place in many business environments, and again security needs to be considered, both the security of the platform and what information is being shared and the recording of the conferencing sessions.
- *Fax:* Today fax machines are very much less in use than they were 20 years ago; however, for those organizations still using such technology, care needs to be taken because sending information by fax machines can sometimes cause security problems:

(a) Makes sure steps are taken to ensure that faxes are indeed sent to the right destination; often has been the case that the sender has typed in the wrong telephone number or the redial button has been hit and the fax arrives at the wrong place.

(b) If sending sensitive business information, also make sure the fax machine is physically secure, that an authorized person is attending the machine as the fax comes through and that the person telephones or emails to confirm that the fax has been securely received and the correct number of pages has come through.

- *Post and courier services:* Sending information by post or by courier services, there needs to be a clear policy stating what types of sensitive and/or critical information can be sent and the means and methods for doing this:

(a) Information at the highest level of sensitivity should use the technique of double-enveloping with the outer envelope not indicating the classification of the contents but the inner enveloped marked.

(b) It is generally accepted practice that envelopes containing restricted information could be sent in a single envelope as long as it does not indicate the classification marking but is marked with the words "to be opened by the addressee only."

(c) Use tamper-proofed containers or boxes or such containers which make it obvious that an attempt at access has been made.

(d) Use registered or recorded mail services.

(e) Use trustworthy and accredited couriers.

6.4.2.6 Information Destruction or Deletion

- *Erasure and deletion of information from IT systems:* Once information has gone past its "use-by" date and no longer needs to be kept, it should be deleted and/or destroyed from the system:

  (a) Electronic files and folders on personal computers (PCs), laptops and servers should be deleted; this can be done using the standard O/S commands or application functions or by using an appropriate piece of software technology that wipes clean the information.

  (b) A word of caution: in most systems, although they have a delete command, the information is not actually deleted; it is only the link to the directory system that is deleted or disconnected, which means it still is possible to recover part or all of the information if you know how to locate it in the system. There are ways of improving this situation other than by physical destruction of the storage media.

  (c) Deletion of information should also apply to all copies as well as backups of the information.

  (d) Having an orderly file directory system helps to facilitate the search for files and any copies or backups.

  (e) The asset inventory should be revised to record the deletion of information.

- *Disposal of hardware and storage media:* Disposal of devices and storage media always creates the problem of deletion of information contained in memory. The simple delete function offered by most devices and applications does not offer complete deletion, as residue will always remain in the system. Devices and storage media containing confidential and personal information require deletion software that completely deletes all information, leaving no remaining residue.

- *Disposal of paper-based information:* Again, as with electronic information, once paper documents and files have gone past their use-by date and no longer need to be kept, they should be suitably disposed of:

  (a) Paper documents, reports and other paper-based information should be appropriately destroyed using approved shredding or other physical destruction methods and technology (e.g., burning, pulping).

  (b) Destruction of paper documents should be carried out by a trusted person or an approved external organization.

  (c) Copies and backups of all information being disposed of should also be destroyed.

6.4.2.7 Information Backups

- *Identification of information to be backed up:* All information should be backed up one way or another: backup information on staff PCs and laptops or information on servers and on other technologies. Some organizations use cloud platforms and third-party data management services for backup.

- *Method of backup:* How, when and where backups need to be done need to be defined. Backups can be done centrally using a network of one or more servers, servers at a specific department level or locally at a user's computer. The policy should define which information applies to which type of backup. For example, operational information is commonly done via a central network server, whereas locally processed information might be done on someone's PC or external hard drive or memory device. It must not be forgotten that it is not only electronic information that needs to be backed up but also paper-based information systems because all information systems, IT or non-IT, suffer from the risk of physical or environment damage or loss as well as theft.

  How frequent backups should be made should be related to the criticality value or level of the information. Daily overnight backups are quite common in many organizations, especially the large ones. End-of-week backups should be the norm and maybe combined with the daily overnight process. Backups need to be done often and regularly, and the more often, the better, within reasons of practicality. The more time that information is left not backed up, the more risk the organizations run of it being lost or damaged.

- *Protection of backups:* Critical and/or sensitive information should be secured to:
  - (a) Protect its confidentiality and integrity such as the use of cryptographic technology or strong access controls.
  - (b) The backup systems should be physically protected to ensure the integrity and availability of the information.
- *Testing restored backups:* There should be a procedure to make sure the backup process and system are working properly and the backed-up information can be read and is a faithful copy of the original. A good maxim to remember is that your backed-up information is only as good as your ability to restore or recover the information from its backed-up state without it being corrupted or lost in any way. What follows from this is that the system and media you are using for backups need to be of a good quality and the process of restoration needs to work properly. If you cannot restore your valuable business or customer information to its original state, then it was not worth the time in backing it up in the first place. Testing the backup system should be done if there are any changes to the backup process and/or business environment using this process.

  Testing the process should be undertaken on a regular basis whether or not there have been any changes. It is best to find out how good your backup and restoration process is before a major disaster or incident happens rather than find out later that it does not work properly; always be ready and prepared before an incident not after.
- *Storage:* Storage of the backups is another issue that needs to be addressed by the policy. The normal best practice for this is to store the backups off-site somewhere away from where server and processing of the information takes place. The reasons should be quite clear: if such backups are kept on the same site, then they face the same risks to which the originals are subject, for example, damage and loss by office fires, physical destruction, theft and other threats. Hence, backup means to not only back up the information but also back up the information somewhere else.

  Store some backups away from home (or "offsite," to use the jargon) in case your backup data gets stolen or damaged along with the computer it is backing up.
- *Storage media:* The type of media used for backups is also another policy issue. There are many types of media for this purpose including tapes, DVDs or CDs, USB sticks, SD cards and other removable memory devices, as well as removable or external hard drives.

- *Backup records:* Accurate records should be kept of what is backed up, where the backups are kept, date and times and other information that might be necessary for incidents, disasters and system failures, as well as for auditing purposes and other information-handling processes.

6.4.2.8 Information Records Management

- *Records:* Keeping records is an important feature of management systems including ISO/IEC 27001. Records provide a vital link between past events and activities, and the effectiveness and efficiency of the ISMS to preserve information security and to manage the organization's risks.
- *Retaining or archiving information:* Certain documents an organization has will need to be kept for a certain period of time as required by law such as tax records and company accounts. For documents outside of the legal and regulatory category, an organization may have both business and technical reasons for the retention of information such as:

  (a) Protect its intellectual property rights (IPR), trade secrets and research results.

  (b) Audit trails, system and network configuration files and other records, especially if they are required to be used as part of an incident handling or a forensic and/or criminal investigation.

6.4.3 Access Control Policy, Procedures and Process

6.4.3.1 Basic Idea

Access control is a very broad subject and encompasses various types of methods and means of giving access to information and resources. For example having:

- The rights and means of access to enter buildings, offices and computer rooms;
- The right and means of access to action and/or process information: obtain, retrieve, copy, store or destroy to prevent unauthorized access or accidental damage to or compromise of information;
- The right and means of access to use or deploy resources, technology, devices, networks and applications.

Access is given to information and organizational resources to enable staff to carry out their particular job function. Hence, access should not be given to specific information if it is not necessary for the conduct of the staff's official business duties. Normally, the principle of least privilege is used to grant users or staff permissions that they need to do their jobs, thus not providing additional permissions that are not necessary to minimize the risk of compromise. This principle should not be mistaken for the other principle of segregation or separation of duties, even though this latter principle employs the idea of giving limited or least privileges.

This does not mean compromises will not occur with those users who have been given access permissions and rights, as there is always the case of those authorized to have access doing damage to the system and its contents. This can be the case with insider trading, staff involved in internal fraud, theft by staff and other incidents where the culprits did have authorization. Hence, ISMS measures need to protect against those not having sufficient access rights and permissions and those owners of the information to control those who do.

Although access control is primarily about ingress, the rights, permissions and privileges to gain entry, there is also a dual protection aspect and that is the egress situation where exiting from a building or system does not leave it in a condition that could be compromised.

It is important from the beginning to have a clear understanding of the business requirements for access: access to what and by whom and the means of gaining access. What are the requirements for access to the different information systems that the organization has? Are there requirements for access to different applications and processes? What are the physical access requirements: staff, contractors, maintenance engineers or visitors? How is access control to be implemented with regard to IT-based information systems?

At a management system level, it is normal policy and good security practice for the information (system or process) to have an owner. This owner is responsible for making sure that the information for which they are responsible as the owner is adequately and suitably protected. At the technical implementation level, access to particular models of access include Mandatory Access Control (MAC), Discretionary Access Control (DAC), Role-Based Access Control (RBAC), Rule-Based Access Control (RuBAC) or Attribute-Based Access Control (ABAC).

6.4.3.2 Access Rights and Information Classifications

The organization needs to decide what access rights are given to information and by whom. Some may be given access to information at different levels of classification. Staff will be given access to information that is on a need-to-know basis.

The higher the classification (the most sensitive information), the smaller the group that have need-to-know rights; hence, the biggest group of users would those having access to restricted information (less sensitive information).

6.4.3.3 Access Control Methods

There are a range of methods that can be used for controlling access, whether relating to physical systems and facilities, IT systems or telecoms systems.

6.4.3.4 Authorization, Accountability and Ownership

At a management system level, it is normal policy and good security practice for information (system or process) to have an owner. These owners are responsible for making sure that the information is adequately and suitably protected. Thus, the owners are accountable for ensuring they have appropriately classified the information, they have compiled an information asset inventory and they will ensure the necessary and appropriate control measures and actions are taken to protect the information at the defined information classification. They need to ensure that such protection is implemented in order to manage the risks to this information. The technical implementation of access control is outside of the scope of ISO/IEC 27001. Authorization for access is also another management aspect of ownership of information: authorization granting or denying access and the appropriate access rights.

The information owner might be, for example, an individual or a head of a group, unit, department or division of an organization.

6.4.3.5 User Access Registration and Deregistration Process

The implementation of access control does have many variants depending on the type of access, the methods of access and the technology being used. However, there are some common elements which should be considered when producing access control procedures.

The overall process should start by registering the users onto the system and then setting up an account. Once this has been established, the users can use the account to access resources on the system according to the rights and permissions that they have been given. This also applies to the access of services, for example, telecommunication services and internet services.

The management and maintenance of access policy and procedures, the rights, permissions and privileges, the control measures and methods for access are not necessarily trivial and can sometimes prove to be quite

Case Study 6.1: Example Information Classification and Access Rights

Case Study-Marking	Rights
Strictest in Confidence	Information available on a strict need-to-know basis (e.g., available only to the CEO and the board of the company).
Confidential	Information available on a need-to-know basis (e.g., available only to senior management, or available to a group in the organization with a specific job function such as finance, human resources, research and development).
Restricted	Information that is not available to the public but is generally internal to all employees or staff.
Public	Information available to the public (e.g., information about a company on its website).

In addition, information can be further classified, for example, a confidential marking might have a caveat attached to it such as the following.

Marking	Rights
In Confidence	Need-to-know rights covering information without a caveat.
Medical in Confidence	Need-to-know rights with the caveat that medical information is restricted to those working in the healthcare field: doctors and nurses.
Personnel in Confidence	Need-to-know rights with the caveat that personnel information restricted to those working in human resources.
Financial in Confidence	Need-to-know rights with the caveat that financial information is normally restricted to the finance department/group, the CEO and directors.

In addition to need-to-know rights, which are a means of allowing users to read information, there are also need-to-modify rights, which give users permission to modify, delete and/or generally create information.

An example of these different types of access is that referred to as read-and-write access. There are other access rights, such the right to be able to execute a particular piece of software code. The following is a list of these rights found in applications and form part of MAC and DAC policies. Some applications do not offer all these options or restrictions.

Access Right	Access: Subject → Object
Read (R) access	The subject can: Read the contents of a file, and list directory contents.
Write (W) access	The subject can change the contents of the information contained in a file or directory with these tasks: Add, Create, Delete, Rename.
Print (P) access	The subject can print a copy of the file.
Copy (C) access	The subject can make a copy of the file.
Execute (X) access	If the file is any software program or executable code, the subject can cause the program to be installed and run.

an administrative burden and have high overhead, if not managed correctly, especially in large organizations with thousands of users and system accounts.

Systematic, properly implemented ISMS processes and the effective deployment of the ISMS measures can reduce this workload. From the point when a user is first registered and given an account on the system, until the time when that account is cancelled and the user's access rights need to be revoked, there needs to be management measures in place to ensure the access process is secure, effective and robust. This process involves a number of measures, both nontechnical and technical, all of which should work in harmony and without compromise.

6.4.3.6 User Access Termination Process

Once employment has finished or has been terminated, or there has been any change of employment, a number of internal security-related activities need to take place and be enforced with regard to user access (see also Section 6.4.4). This applies to the termination of the employment of full-time staff as well as contractors. Such activities include, but are not limited to:

- Return of organization assets (for example, IT equipment, mobile phones and other devices belonging to the organization, information, documents, files);
- Removal of access rights and privileges;
- Enforcement of ongoing legal responsibilities contained in confidentiality agreements and terms and conditions of employment continuing for a defined period after termination.

Requirements with regard to the termination of employment related to security aspects should be addressed in the contract of employment. Similarly, with third-party contractors and agency staff, security requirements with regard to termination should be included in their contracts.

6.4.4 Acceptable Use of Resources

6.4.4.1 Basic Idea

Many organizations do allow their staff to have access to the internet for personal use as a goodwill gesture to improve employee relations. In such cases, as well as the general use of the organization's assets and resources for business, it is a very good idea to put in place acceptable use policy with regard to the use of the organization's resources, in particular, its IT systems that are deployed for email, internet access and services.

The basic intent for such a policy would be to protect the organization, its staff and management, its business partners, customers and suppliers from individuals involved in any illegal, incriminating and/or damaging actions, whether knowingly or unknowingly. The consequences of such actions could expose the organization and its staff, its business partners, customers and suppliers to a range of risks, impacts and possibly civil and criminal legal proceedings, actions and lawsuits.

The following is a word of caution regarding the content of the Acceptable Use Policy. An organization that does not explicitly state what is not acceptable and what is prohibited might face some staff relations issues. For example, a member of staff is dismissed for unacceptable use of the internet for personal use. The organization faces the risk that the member of staff files a lawsuit for unfair dismissal and is claiming compensation because the activities in which they were engaged were not expressly prohibited. It is always good practice to make it clear to staff that access to the internet and email facilities for personal use is a privilege and not a right.

6.4.4.2 General Policy Provisions

In producing this type of policy, the following are some notes that should be considered. The policy should:

- Apply to all levels of staff, as well as contractors, consultants and other individuals whom the organization might employ on a short-term or temporary basis.
- Provide a clear definition of what is considered personal use including information on how much access time is acceptable and when personal access is allowed, as well as expressly stating that the user should exercise good judgement regarding the reasonableness of personal use.
- Provide a clear warning of what not is allowed and strictly forbidden and prohibited, including the following:
  - Accessing pornographic or indecent websites and downloading material from these websites;
  - Using chat rooms, social media or online messaging services in particular for the exchange of information, comments or opinions that are sexist or racist or any of these that will cause offence, including the use of offensive and indecent language;
  - Any activities or transactions that are fraudulent or meant to deceive or where the user masquerades as another user;
  - Any activities or behaviour that might provide unauthorized access to other organizations' information, their systems and the network services they use and cause, for example, disrup-

tion or denial of services or damage to or loss of information and processing resources.

- Provide warnings with regard to the need to comply with any copyright, digital rights and licensing requirements and restrictions related to material on websites.
- Provide instructions with regard to downloading material from the internet including warning of the cautionary measures to be taken to avoid downloading malicious code such as viruses, worms and Trojan horses and executable programmes.
- Provide a clear statement of what disciplinary actions are to be taken if staff do not comply with the policy.
- State what the organization's rules are with regard to its right to audit and monitor networks and systems on a regular or periodic basis. Staff should be advised that their access to the internet and use of email may be monitored in accordance with the laws in whatever jurisdiction the organization operates.

A note of what this might include can be a useful reminder to staff as a cautionary measure. For example, the fact that websites that have been used can be traced, the frequency and time spent at sites will appear in an audit trail and search history of using the internet resources for personal transaction (online banking, bookings and orders).

6.4.4.3 Specific Policy Provisions

These might include:

- Warning users not to reveal their account password to others or allowing use of their account by others, including those outside of the organization such as family or friends in the case of a laptop being used off-site;
- Warning users to make statements with regard to warranties, licences or any other statements that might legally commit the organization, expressly or implied, unless it is a part of the user's normal job duties and prior authorization to carry out this action has been given;
- Reflecting the organization's requirements for protecting the information that staff create on their systems and what is deemed to be the acceptable (and legal) use of encryption for information that is considered sensitive or especially vulnerable;

- Monitoring and/or intercepting a communication network or service for the purpose of intercepting data unless this activity is a part of their job function and prior authorization has been given and it is allowed in the jurisdictions covered by these activities. For example, the UK Regulatory Investigatory Powers Act (RIPA) law covers some such activities.

6.4.5 Human Resource Policy, Procedures and Process

6.4.5.1 Basic Idea

An organization's human resource (or personnel) group is generally there to recruit, develop and utilize its personnel for the purposes of meeting the organization's operational and business needs.

It has been shown many times in surveys and through work experiences that the biggest threat to information systems originates from people. The people threat could be accidental (e.g., errors in entering customer information into an ordering system or the unintentional deletion of files on a PC) or intentional (e.g., the abuse of organizational ICT systems for personal gains or destroying critical data as might be the case with a disgruntled employee).

Therefore, management should focus its effort on getting the people aspect of information security right. This involves several things (Tables 6.1 to 6.3):

- Achieving an internal security culture;
- Awareness and training;
- Human resource management support;
- Regular updates on security issues;
- Getting staff involved in discussions, user feedback and reviews.

6.4.5.2 Training and Awareness

Information security training needs not just to achieve knowledge transfer and understanding security policies and procedures but should also aim at getting staff involvement and commitment to adopting the procedures. Just knowing the procedures without practice does not modify the risk of loss or damage to the organization's information. Some specific information for security training and awareness aspects (the content of which should be, at a minimum: the information security and business risks, rules and guidelines, roles and responsibilities and commercial and legal aspects) include:

Table 6.1
Recruitment Activities

Activity	Commentary
Recruitment	The human resource group is responsible for facilitating the recruitment of the right people to fit job vacancies. Defining employee skills, qualities and competence is an organizational determined aspect. In the area of information security, there are some specialist jobs such as a security officer/manager, risk manager or IT security expert. In addition, there will be staff that are not specialists but will need to know about, and how to use, information security policies and procedures in the day-to-day work.
Screening	This process involves assessing the applicant's suitability for the job advertised. This normally involves screening applications based on their work experience, skills and training for the described job. This normally involves a two/three-stage process. Screening at a general level for example reviewing applications, requesting a CV, then this could be followed by getting letters of reference from previous employers and from educational establishments, certificates as proof of their qualifications. Depending on the type of job being applied for, a further screening stage might need to be carried out. For example, for some security-specific jobs, there may need to more detailed assessment of their previous work and/or social background. This could go as far as checking whether the individual has a criminal record or checking their financial state by doing a credit reference check.
Interviews	Like most recruitment drives, the interview activity is a critical part of the process involving the human resource group, representatives from the department involved in the vacancies and the candidates themselves. The more detailed screening process normally comes after the interview relating to those that have been sort are listed.
Contracts and nondisclosure agreements (NDAs)	The new employee will be expected to sign a contract of employment with the employer's terms and conditions. They will also be asked to sign an NDA or confidentiality agreement, as well as various forms for their salary payments, and whatever security forms are deemed necessary, for example, an acceptable use policy.

- Processing of personal data and/or the organization's confidential and critical information;
- Information-processing activities;
- Use of IT systems and software, the internet and email, and mobile and Wi-Fi services;
- Reporting incidents, system malfunctions and failures;
- Backup of data;
- Physical protection;
- Dangers of social engineering.

Table 6.2
During Employment Activities

Activity	Commentary
Employee responsibilities	Staff will be assigned certain roles, responsibilities and duties during their employment, which are related to the organization's information security policies and procedures. These include general security duties and responsibilities as well as more specific security duties and responsibilities.
Employee system accounts	During employment, users would normally be given a system account to allow them to access the organization's IT systems and networks. What they can access and what rights they have to read, write, print, copy or execute files on the system will largely depend on their specific job role. It is good practice for the human resources group to keep a record of this on their system.
Provision of training	Staff needs to be given training according to their jobs function and needs. The objective of this should be to provide: • Awareness of information security and risks; • Instruction on how to use security-related policies and procedures; • For career development. There are various forms training can take such as on-the-job training, formal training courses, online training and at-the-desk self-instruction using training software.
Review of training	As business needs change, staff development and/or staff get moved to different job functions and/or responsibilities, then training needs will likely change. This will entail reviewing staff training, providing additional or new training and updating the staff records. It is important that staff keep up-to-date with the organization's developments in information security.
Disciplinary process	Security compromises involving staff can and do disciplinary action depending on the severity of the incident.
Legal representation	Some information security incidents can involve civil or criminal activities. Members of the staff who are named as being the perpetrators or being involved in any way could face legal action against them as individuals and possible action against the organization itself.

6.5 Sector and Application Specific Controls

There are a number of sector specific ISO/IEC standards that have been developed that specify controls and implementation guidance for individual sectors. These include:

- ITU-T X.1051|ISO/IEC 27011—Information Security Management Guidelines for Telecommunications Organizations Based on ISO/IEC 27002 Telecommunication Organizations;
- ITU-T X.1631|ISO/IEC 27017—Guidelines on Information Security Controls for the Use of Cloud Computing Services Based on ISO/IEC 27002;

Table 6.3
Termination of Employment Activities

Activity	Commentary
Termination responsibilities	Staff employment gets terminated for various reasons: (1) voluntary termination say if the member of staff wants to move or they need to move for domestic reasons, or (2) involuntary termination such as being fired or sacked for disciplinary reasons or maybe they are made redundant. For these different types of circumstances, the responsibility of the handling the termination of staff employment should be assigned to someone in the organization's human resources group with the appropriate seniority level. This role should involve making sure the process is handled correctly to ensure that all business, legal and personal issues are concluded in a proper and effective way appropriate to the circumstances of the termination. Some terminations can be the result of disciplinary action and a severely disgruntled employee with bitter feelings towards the organization. These types of situations need to be handled sensitively and extreme caution cognizant as to what damage the employee might cause before leaving. In some organizations, employees are asked to the leave the premises immediately without serving the normal period of notice if there is a risk of a security compromise.
Return of assets	Staff is normally expected to return all assets that belong to the organization. This should especially be the case with all computing equipment, storage devices, software and applications as well as paper documents, files, reports, manuals and/or procedures containing the organization's information, in particular that which is sensitive, confidential, restricted or related to personnel.
Removal of access	To avoid the risk of unauthorized access to buildings, offices and rooms, pass cards, identity cards and other devices should be collected from the member of staff. Also any physical entry personal identification number (PIN) codes should be removed from the system. All access accounts to the information systems, in the name of the member of staff, should be removed from the system. This is to avoid unauthorized access by the former member of staff or a colleague or accomplice at a later date. This applies to all ICT equipment from system servers, PCs, laptops, mobile service accounts and any other device, which allows a user to gain access to the organization's information.

- ISO/IEC 27018—Code of Practice for PII Protection in Public Clouds Acting as PII Processors;
- ISO/IEC 27019—Information Security Management Guidelines Based on ISO/IEC 27002 for Process Control Systems Specific to the Energy Utility Industry;
- ISO/IEC 27799—Health Informatics—Information Security Management in Health Using ISO/IEC 27002.

These standards provide organizations with a source of sector controls they can use during the risk treatment phase, that is, when determining a set of controls that could be considered (ISO/IEC 27001 Clause 6.1.3 (b)).

6.6 ISO/IEC 27001 Annex A

6.6.1 Purpose and Definition

Annex A is solely a reference set of control objectives based on ISO/IEC 27002. It is important to note that Annex A itself, or any of its content, is not a requirement. The prime purpose of Annex A of ISO/IEC 27001 is to support requirements of ISO/IEC 27001 Clause 6.1.3 (c) and, in this case, it is checking the set of controls that organization has determined as necessary against a reference set to see whether any control areas or controls have been overlooked. This is merely a type of gap analysis exercise and is not a replacement for the determination of control exercises. Therefore, it is incorrect to say that, to conform to ISO/IEC 27001, an organization must implement the controls in Annex A. It is correct is to say that, to conform with ISO/IEC 27001, the organization must fulfil the requirement of Clause 6.1.3 (b) and determine its own set of controls (Clause 6.1.3 (b)) and then fulfil the requirement of Clause 6.1.3 (c) and carry out a comparison with the reference set of controls in Annex A. It is therefore incorrect to say the controls in Annex A are mandatory and are requirements to be implemented.

Even if there are controls in Annex A that have not been determined as necessary, this does not mean that they need to be added to the list of controls to be implemented. However, there may be controls that have been overlooked during the determination stage and consideration should be given as to whether they should be implemented. In all cases, any controls that are not included in the implementation list but appear in Annex A do need some justification as to why they are not included.

Figure 6.2 shows the two processes: (1) determination of controls to be implemented, and (2) benchmarking comparison against Annex A controls.

The controls listed in Annex A of the third edition (2022) of ISO/IEC 27001 are based on the third edition (2022) of ISO/IEC 27002. There have been substantial changes made to these controls compared to those in the second edition (2013) of ISO/IEC 27002. It is important to note that only the control statements from ISO/IEC 27002 are listed in Annex A. There is no control guidance, control attributes or other information in Annex A, as such details are not relevant to implementing or conforming with ISO/IEC 27001. Conformance with ISO/IEC 27001 does not include implementing any of the guidance or attributes given in ISO/IEC 27002.

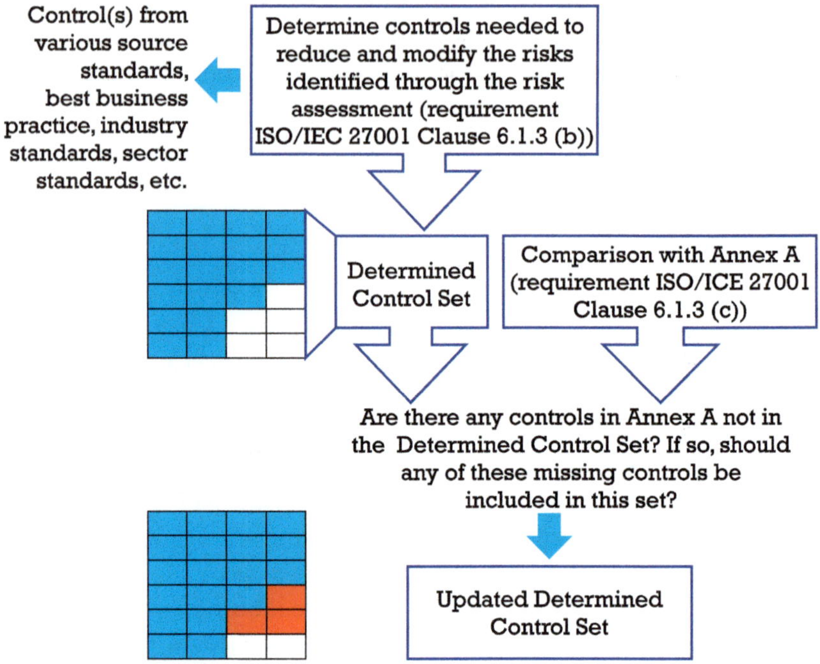

Figure 6.2 Annex A in ISO/IEC 27001:2022.

The controls in Annex A are very generic in nature and their adoption would need to be customized to meet an organization's requirements. The Annex A controls are organized into four categories:

- Organizational controls, for example, dealing with policies for information security, roles and responsibilities (management and staff); asset management, access control, identity and authentication management, access rights, information sharing, classification and labelling, threat intelligence, security aspects of supplier agreements, supply chain management, incident management, ICT readiness, protection of records, protection of PII, legal requirements, review of information security and operating procedures;
- People controls, for example, dealing employment (recruitment, contract terms and conditions, screening and vetting, responsibilities, discipline and termination), NDAs, training and awareness, remote working or security incident/event reporting;
- Physical controls, for example, dealing with the physical protection of equipment, offices, information physical protection, storage media,

offices and buildings, maintenance of equipment and physical monitoring;

- Technological controls, for example, technical aspects of access control, network security, malware, use of cryptography, vulnerability management, application security, backups, monitoring, software installation, security testing in development, secure system architecture controls, outsourcing, change management and configuration management.

Not all of the controls in Annex A are applicable to every organization. The controls that are applicable are those that are required to address the specific needs of the organization and the risks that have been identified. However, first and foremost, the organization shall focus on determining controls as per Clause 6.1.3 (b), and only then should it turn to Annex A in conformance with Clause 6.1.3 (c) to check whether there are any controls are missing from the set of determined controls.

CHAPTER

7

Contents

7.1 Operational Planning, Management and Control

7.2 Ongoing Risk Assessment

7.3 Ongoing Risk Treatment

7.4 Example Sketches of Operational Threats

7.5 Examples of Sketches of Operational Processes

7.6 Incident Management Narrative

7.7 ISMS Availability and Business Continuity Narrative

7.8 ISMS Use Examples

ISMS Operations

It is not in our control to have everything turn out exactly as we want, but it is in our control to control how we respond to what happens.

—Accredited to a Stoic philosopher

7.1 Operational Planning, Management and Control

7.1.1 Processes

There are several types of processes that an organization needs to have in place to manage, control and protect its day-to-day operations and its information. The processes in ISO/IEC 27001 (Clause 8) are generic in nature and will need to be adapted to include specific features unique to the business context, objectives and operations. These processes are a set of actions designed to achieve effective ISMS performance within its defined scope. The process design is nonspecific, and it is up to the organization to decide how they will implement the process; for example, the ISO/IEC 27001 risk assessment process leaves it up to the organization to select which assessment method or technique to use. These processes are designed to assist and ensure that the ISMS can achieve its desired outcomes to prevent or reduce the risks and any undesirable effects and impacts caused by risks and achieve continual improvement and at a level of information security commensurate with the level of identified risk.

The organization needs to have appropriate processes in place in accordance with the requirements of ISO/IEC 27001. The ISMS should be integrated with the other processes being implemented by the organization and that information security is implemented alongside other organizational processes, controls and management structure.

The processes need to be documented to provide information that the actions as planned have been undertaken, for example, documented proof that the risk assessment process has been undertaken and the results are recorded in a risk assessment report; a document giving the risk treatment to be implemented; and a record of what actions need to be taken and when these actions have been completed, any problems with the implementation and any variances or divergences from the treatment plan.

Changes are inevitable within a business and external to the business: some of these will be planned changes and others will not be planned resulting from unintended, unanticipated circumstances or the consequences of accidental or chance incidents. Always expect the unexpected to happen, changes to happen and unwanted incidents to occur. These unintended changes and their consequences need to be managed and the planned changes need to be managed and controlled. All changes need to be reviewed as they might alter the risk profile of the organization, which means that the controls and measures to mitigate against the risks may need to be changed.

7.1.2 Example Processes

There may be operational processes that are common across the organization and some that may be specific to certain departments, business operations or services. In ISO/IEC 27001, some processes are clearly identified, such as risk assessment, risk treatment and internal audits. Other processes are not clearly defined but come under the general heading of an information security process, that is, they support a control that has been determined during the risk treatment process (ISO/IEC 27001 Clause 6.1.3 (b)), for example, information security incident management process, business continuity and disaster recovery processes, through the processes for the use and management of technology to handle, store and communicate information in a secure way.

7.1.3 Training, Awareness and Usage

It is essential that the organization's management not only makes employees aware of the procedures that are applicable to the job role and function, but also arranges for employees to have some form of training in the use of these procedures. A new recruit could easily compromise the information security of the organization by making errors or mistakes or doing the wrong things by a lack of understanding and knowledge about what is the correct security procedures to follow in their day-to-day job. This may not just happen to new recruits, but employees who have been in the organization for a while may need to use a new procedure or a revised version of an existing procedure. In all cases, the same applies; a lack of understanding and know-how can lead to an incident and hence compromise the information security of the organization.

It is also important that all employees use the procedures provided by the organization to ensure consistency and the intended performance. If employees use their own methods of doing things in an operational environment, then this could not only compromise the information security of the organization and its systems, processes and services, but it would be against the organization's policy and rules and could be rewarded with disciplinary action. Finally, employees using nonstandard procedures are likely to result in a nonconformity being given in an audit. We should remember that a well-written procedure would be generally designed to achieve the most effective level of information security and so the instructions should be carefully thought out and written in a way that will achieve this level. A nonstandard procedure used by some employee may not only fail to achieve this level but would be against the organization's policy and wishes.

7.2 Ongoing Risk Assessment

A risk assessment is always just a snapshot in time based on the perceived risks and impacts at the time the assessment was carried out. There should be no doubt that business conditions will change and, with this, the risks and impacts related to the organization will change. Therefore, it is imperative that the organization keeps up-to-date with its risk profile by reassessing its risks and impacts on a regular basis as well as updating its risk treatment following this risk assessment.

What a regular time is depends on several factors, including:

- Dynamic nature of the organization: operating in a case in which it is most likely subject to frequent changes in business conditions: vitality of the market, its growth rate, its diversity and other variable conditions.

- The complexity of the organization: the more complex it is in terms of processing resources, the volumes, the type and nature of the information involved and the number of interfaces and dependencies it needs to manage and take account of, the more risk factors and risks with which it may need to deal.
- If the nature of the organization's business places it in the front line for attack, then the frequency of attack would be higher.

There is no set standard time for reassessing its risks and impacts. Some organizations do a reassessment every 3 to 6 months; some leave it to 9 months; the actual frequency can only be determined by the organization itself with management making the decision on the frequency. In addition to a regular programme of risk assessments, then there will always be the chance that a risk assessment needs to be carried out sooner than the next scheduled risk assessment. This might be after a major information security incident, a major system disaster, a significant change in the organization circumstances or after a management review or an internal ISMS audit (ISO/IEC 27001, Clause 9). A reassessment of the risks may be needed as the result of an external certification audit.

An organization should implement a change management process (see Section 7.4.4) that includes information security aspects.

7.3 Ongoing Risk Treatment

As with risk assessment, risk treatment is also an ongoing process. In order to keep up-to-date with changing conditions and circumstances that have risk implications, the organization must be continually alert and be monitoring the business situation. The risk environment can change gradually over time or accelerate within a relatively short time; the organization needs to be prepared for both and be able to identify, assess and find a treatment for the risks.

There is no set standard time for reassessing its risks and impacts. Some organizations do a reassessment every 3 to 6 months; some leave it to 9 months; the actual frequency can only be determined by the organization and the nature of its business operations.

It is important that the organization not only produces a risk treatment plan, but it implements this plan in a timely manner and it is implemented appropriately to ensure optimal effectiveness.

7.4 Example Sketches of Operational Threats

7.4.1 Malware

7.4.1.1 General

Malware attacks are a worldwide headline-catching piece of news. The virus is an example of a piece of malware or malicious software, which is designed to infiltrate or damage a computer system. This class of software also includes worms, Trojan horses, spyware or adware, to name a few pieces of malicious software code.

Malicious code is one case of those pieces of software that is designed specifically to do damage and harm and put organizations infected with such code at risk. As is well known by many cases, viruses, once in the system, attach themselves to applications software and can then replicate and spread themselves very fast like a fire in a forest that has not seen rain for weeks or months. In their path through the system, they cause destruction and damage just like the fire and, even when the virus has been discovered, it may not be immediately eradicated, just like a fire that has taken hold of the forest and might take some time to extinguish. Therefore, for a period of time, the virus is out of control. The virus might be deleting files and damaging software programs and applications; they may even be aimed at reformatting the hard disk. The more benign types of viruses can also cause the organization problems because continual replication of the virus can eat up memory, trigger erratic behaviour and maybe make the system crash, all resulting in degradation of performance and a possible denial of service.

The worm is a self-replicating piece of software or code, which does not attach itself to an application, but nevertheless causes widespread damage across networks such as the internet. Because it is self-replicating, it, like the benign virus, eats up network bandwidth, consuming the computer system's memory, slows down systems and, in the worst case, causes a widespread denial-of-service attack. Worms such as the Morris Worm of 1988 caused global disruption and chaos to many systems connected to the internet at the time once the worm had invaded these systems. Several of the worms that are released onto the internet only replicate and spread themselves; however, some worms have been produced to carry payloads such as to delete files, send files via email to other systems, install back doors in systems and even encrypt files. The backdoor problem can result in an opportunity for the attacker to create a zombie computer within the infected system, which, when similar zombie computers are networked together to form a botnet, subsequently can be used by those responsible for sending spam, which, in turn, can create several security incidents as discussed previously.

The Trojan horse is a piece of software code that contains and/or installs a piece of malicious code into a system. This piece of malicious code is often called a payload or simply a Trojan. These Trojan horses might appear as useful and harmless pieces of software, but beneath the surface lurks the malicious code, which, when the seemingly useful piece of software is executed, triggers a sequence of damaging events. The Trojan horse can cause unauthorized activities to occur on the victim's systems without his or her knowledge or consent such as:

- Downloading and uploading files from the internet;
- Sending out emails using the victim's email address book including the distribution of spam;
- Phishing in the victim's files to gain access to credit and banking details;
- Causing the system to restart;
- Attaching the system to a botnet;
- Deletion and corruption of files.

Trojan horses can infect systems in several ways such as delivery as an email attachment or by a user downloading a file from a rogue site on the internet.

7.4.1.2 Managing the Problem

Protection from malware needs proper management processes and controls in place. For a start, the organization should produce a policy prohibiting the use of unauthorized software and then implement controls to prevent or detect the use of such software. The organization should also have a policy regarding the downloading or receiving of files and software.

A set of procedures should be produced by the organization of how to deal with and report malware problems. Responsibilities for dealing with malware should also be defined and this should be included in these procedures. Staff should be trained in how to deal with malware.

The organization should conduct regular audits on the software that supports its critical systems and processes: such exercises should search for the existence of any unapproved files or software.

The organization should implement software that is capable of detecting malware on computer systems and media. This type of software should be capable of scanning files, email attachments or web pages for the presence of malware infections. There are a number of commercial products on the market for this purpose. The functional capability of these products does vary generally; they are designed to detect, prevent and eradicate malware from computer systems and other IT devices. They generally scan the system files and folders located on hard drives and other memory devices. The malware software relies on a dictionary of known pieces of malware to compare against when scanning the system. The effectiveness of this software relies on keeping the dictionary up-to-date: most product vendors have an online facility for getting live updates. This live update needs to be done on a regular basis, at least once a week if not several times a week. It is a good idea to set your system to do an automatic update.

As a word of caution, the malware code writers always try to stay one step ahead of the vendors by using various techniques to disguise the malware to circumvent the dictionary match process.

Payloads such as those attached to Trojan horse are more difficult to protect against. Most of these are delivered by email and there are some email best practice measures that could help:

- Be alert to emails that are received from someone that the user does not know or suspicious or known attachments.
- Set the email preferences to not automatically open attachments.
- Update the antivirus software automatically.
- Update the operating system with the latest patches.

7.4.2 Unauthorized Access

7.4.2.1 General

Unauthorized access to information can lead to disclosure, modification or unauthorized use of the information. Such access might happen as a result of an internal compromise of organizational policy and abuse of privilege, or it might occur as a result of someone externally breaking through the organization's security system. Unauthorized access to information might occur by breaking through layers of physical protection or electronic protection or a combination of both. Access via the internet or other network is a common problem. Whatever the means and method of access, once the unauthorized person has gained access to the organization's information, then he or she has access to an important asset of the organization and an opportunity to steal, modify or destroy this information.

Depending on the type of information being accessed (sensitive, critical or personal), then the organization may be facing a range of problems. For example, if the unauthorized person has gained access to customer details including passwords, PINs and bank account details and sells or publishes these details online, then the organization faces commercial and legal problems. Another example might be that the unauthorized person who gains access to critical information that controls an industrial process and the same person modifies the information or takes control of the process, then this could be a serious safety problem as well as a potential commercial and legal problem.

An unauthorized access attempt might be from outsiders (e.g., hackers, competitors, industrial spies) trying to break into the organization's system and gain access or it might be an attempt by insiders abusing their privileges or using their inside knowledge of the organization to gain access to areas of the organization to which they have not been given access.

7.4.2.2 Managing the Problem

The problem of unauthorized access to information can be managed at different levels. The organization should have an access control policy; it should also have specific procedures for different access control methods and covering different types of access: access to information, applications, processes, IT and networks. The control methods implemented to manage access should be strong enough to prevent unauthorized access. Control of these methods and the information used for gaining access needs proper management and administration. People controls need to be in place relating to who has access, for what purpose and for what information access is given, via what devices can access be made: establishing authorized computer, system and network accounts, password and privilege management, access restrictions, reviews of access rights, disciplinary action regarding abuse of rights and training of users on access control methods, rights and privileges.

7.4.3 Insider Threat

7.4.3.1 General

The insider threat is a malicious threat to an organization, which is becoming more a common occurrence. Such a threat could be from current staff and employees from within the organization, but it might also be from former employees or even contractors or business associates who are working closely with the organization and are given certain internal access privileges because of the nature of the work and collaborative ventures in which they are involved.

The insiders, employees or staff members, will have computer accounts given to them for legitimate, work-related purposes to undertake their job function and work-related duties. This means they have authorized access to computer systems to carry out certain job-related tasks unlike outsiders, who do not have such access (the case of the on-site contractor is different). The insiders may abuse or misuse these privileges and cause harm or damage to the organization. In addition, the insiders generally have far more knowledge, experience and familiarity with the workings of the organization: where things are, who does what, how things operate, the strengths and weaknesses of the organization and the rules and procedures. With this inside knowledge of the organization and the privileges they have, the insiders have much more scope for attacking the organization and being successful than the outsiders, although we should not underestimate or ignore the power of the outside threat. Hence, it easier for the insiders, than the outsiders, to circumvent any information security the organization has in place.

Some of the insider attacks are for personal gain; others are for revenge. Both of these cases are intentional and deliberate attacks. However, an insider threat might be due to user error or mistakes, that is, a threat caused by accidental or unintentional circumstances.

A member of staff might be using the organization's resources for his own personal gain: running his own private business for monetary profit, selling commercial information to the organization's competitors or committing fraudulent transactions, the theft of intellectual property or selling stocks and shares.

Staff members may be disgruntled and unhappy with their working conditions or believe that management or their peers have treated them unfairly. In this case, they might take out their dissatisfaction or anger regarding the situation by damaging system resources, deleting or corrupting information, leaking information, destroying files, deliberately upsetting customers or obstructing and disrupting normal business operations and other acts of sabotage against the organization.

7.4.3.2 Managing the Problem

The insider problem can be managed at different levels, with a combination of including:

- Policies and procedures: include instructions and information regarding insider threat.
- Training: include insider threat awareness into staff training.
- Recruitment process: check employment references, check for past criminal offences or review reports of suspicious or disruptive behaviour.

- Personnel management: manage conflicts, disputes in the work environment; review working conditions of disgruntled staff, provide counseling and dispute resolution opportunities for disgruntled staff or be vigilant to unusual behaviour, unusual work patterns against what the organization defines as standard, work-related activities and patterns of normal work operations.
- Ensure the risk assessment and risk treatment take into account the insider threat problem.
- Termination of employment: implement a secure employee termination process.
- Implement controls for separation of duties and least privilege; enforce strict password practices; secure backup and recovery processes; and implement effective change control procedures.
- Access control: implement strict access controls, enforce strict access restrictions, effective management of privileged users, effective monitoring, logging and audit of system access or restrictions on remote access or review access rights.
- Online activities: be vigilant to online employees' behaviour and manage the use of online services or social networking through the acceptable use policy.
- Enforce an appropriate policy and procedures for BYOD.
- System vulnerabilities: keep the door closed to unauthorized access and exploitation of system vulnerabilities by fixing, repairing or safeguarding against system vulnerabilities.

7.4.4 System Availability

7.4.4.1 General

If the organization wants to keep its business running and to operate productively, then it needs to ensure the resources it needs to do so are available. Availability is one of three information security objectives, along with confidentiality and integrity. Availability relates to all types of resources: people, information, processes, services, networks, IT and facilities. One example is how reliable, for example, the organization's network is in providing for its operational needs. Does the network fail? Does it slow down? Does it provide for 24 × 7 uninterrupted usage? The backbone of most businesses is that of IT. The organization might suffer from a poor record of IT system failures. Why? Is it lack of adequate maintenance or failure to install the latest software upgrades and patches, or is it that the demand for a specific IT system far outweighs its capacity to deliver against these demands?

If an organization's IT system fails or its online presence is interrupted, it could suffer many hours of downtime, which translates into impacts such as loss of revenue, loss of customer confidence and possibly penalties due to a breach of contractual obligations if the loss of availability stops the organization from being able to honour such obligations.

Organizations can improve their chances of having the resources they need being available as and when they need these resources. In particular, it is the resources that provide, process and support its information, as well as the availability of its information that is the main issue.

In the case of IT and network resources, it is a case of regular maintenance and testing of these resources to lower the chances of IT system failures and outages, network service disruptions and protecting these resources from attack, for example, a denial-of-service attack, which can overload, slow down and generally disable systems and their ability to function normally. This is particularly the case with software maintenance, upgrades, patches and vulnerability testing, but can also apply to hardware.

7.4.4.2 Managing the Problem

The availability problem can be managed at different levels, including:

- *Information:* protecting information from unauthorized disclosure or modification, access to information, backups of information, information recovery in the event of a system failure, procedures for handling information, ownership of information assets and authorization to access information.
- *Services:* management of service delivery, service provider contracts and SLAs, managing the connections to services, access and usage of services (e.g., networks, web services, managed security services, disaster recovery services, outsourced services, teleworking services, media disposal) and authorization to access services.
- *IT system usage:* effective access controls, restrictions of usage to avoid misuse, policy for acceptable use and procedures for correct use, secure processing of information using IT, employee terms and conditions on the use of IT, network services and the handling of information and authorization to access IT systems.
- *Software management:* restrictions on software installations and usage, controlling software versions, upgrades and patches, separate private and business use, controlling software support and maintenance,

default or secure configurations, access to utility versus application software, restrictions on remote use of software, control of software licences, software in an operational environment should be kept separate from software in a development and testing environment, malware protection, software backups and secure control over operation software.

- *Vulnerability management:* risk assessment and impact analysis of the vulnerabilities, understanding and identifying system vulnerabilities, assessing and reviewing the vulnerabilities, taking correct or preventative actions to remove or diminish the weaknesses or to improve protection to avoid the chances of exploitation, improvements and controls to manage vulnerabilities need to be tested, ensuring software patches are from an authentic source, the vulnerability process needs to work in collaboration with the incident handling (Section 7.5) and business continuity processes (Section 7.6).

7.4.5 Social Engineering

7.4.5.1 General

Social engineering is a type of information security attack that exploits the psychological vulnerabilities of people by coercing or manipulating them to perform activities, which results in them divulging information: the art of hacking into the human psyche. This type of attack can be performed online (e.g., emails, phishing attacks, social networking) or off-line (e.g., over the phone to obtain information, or posing as officials or service technicians to gain access to company premises, facilities and information). This attack is a confidence trick, but differs from the traditional confidence trick in the way it manipulates the mind of its victim. The aim is to obtain information to gain system access, to steal commercially sensitive information or to commit a fraud or a scam.

There has been a dramatic increase in this type of attack with the widespread use of social media platforms and the ability to entice or tempt users through these platforms to click on links to download malicious and infected files or reveal personal information.

Remaining vigilant if contacted by third parties, check the validity of the source of the communication, email address or message text, and stop, think and check links from unfamiliar sources; do not share personal information. These and many other helpful considerations will build a defence against attacks.

7.4.5.2 Managing the Problem

The social engineering problem can be managed at different levels, including:

- Consideration of social engineering attacks in the risk assessment and risk treatment exercises is a must.
- Identify information as sensitive, critical or personal and evaluate the exposure of this information to social engineering attacks.
- Establish policy and procedures for handling social engineering attacks.
- Train employees to understand the problem of social engineering and the use of social media platforms and to recognize situations that might be an indicator of, or precursor to, a social engineering attack and understand how to apply company policy and procedures to social engineering attacks for protection information.
- Carry regular audits and reviews of the controls to counter social engineering attacks: people, processes, procedures and access control methods.
- Establish roles and responsibilities for handling sensitive, critical or personal information.
- Keep up-to-date with the latest information on social engineering attacks and feed this information back into the training programme and use this information in the revision of policy and procedures.

7.5 Examples of Sketches of Operational Processes

7.5.1 Protecting Information in the Operational Environment

It is important that the organization has the necessary management controls in place to support and secure its operations and its information being processed during operations. Today's multiplicity of technologies for processing, sharing, storing and communicating information adds complications to the task of protecting this information. The management controls can include:

- Procedures for all operational activities associated with the processing of information, in particular for the protection of this information;

- Change management process that covers changes to those business processes, systems and services supporting the processing of information;
- Capacity planning: ensuring adequate resource is available to meet operational requirements;
- Backup process: to protect business and customer information from loss;
- Malware protection: to protect operational systems against malware attacks;
- Separation of business environments: operations versus development and testing;
- Service management and supply-chain management: information security related to service delivery, third-party services, contracts and SLAs;
- Operational monitoring and logs: controls are needed to monitor and record information security-related events;
- Operational software protection: control measures need to protect operational software;
- Vulnerability management: control measures need to be implemented to deal with proper identification, evaluation and resolution of operational system vulnerabilities;
- Operational audits: assessing and verifying the operational environment is adequately and suitably protected.

7.5.2 Backups

7.5.2.1 General

There are two main reasons why backups are required. The first, and most important, is to recover information that becomes lost or corrupted due to some system failure, disaster of compromise. Many users and organizations suffer some sort of loss of information at some point in time; it is a common experience shared by many IT users.

Another reason for backing up is for information retention purposes: to enable copies of information to be brought back into the system that were created and processed at some earlier time.

7.5.2.2 Policy

A policy should be produced which defines the business requirements for backing up the organization's information assets. The slogan "back it up before your business loses it" is a very simple way of illustrating this basic requirement. Performing regular information system backups is vital to ensure an organization's information is adequately protected and preserved from damage or loss. The policy should state the business importance of doing backups and the consequential risks that could severely impact business operations for failing to implement an effective backup process. This should include the risk of exposure to legal proceedings, not being able to respond to customer orders and supplier payments, deliveries cancelled or production lines being halted or running inefficiently. Alternatively, having a good backup system enables the organization to operate normally to continue trading and supplying customers with services and/or products even if the original information has been damaged or lost.

7.5.2.3 Roles and Responsibilities

Information owners are accountable for its safe and secure protection. The deployment of risk controls to provide such protection may be delegated to others for practical purposes, but the owner still remains accountable. It should be made clear that it is essential that all information that is processed by all departments, operational groups and staff, whether they are owners or custodians of information assets, should be the subject of regular backups.

Much of the information an organization has and processes is indispensable for it to do its business. Also, the organization needs to take the responsibility for all the information it handles, even though the information owner is another organization or its customers/clients. The consequences of loss information or its destruction or damage can have severe commercial and financial consequences, as well as dire legal implications, so back it up before your business loses it is a clear and simple message.

7.5.2.4 The Technology Is Available; Use It!

Backups can be implemented in many different ways, and today's technology makes the process relatively easy. There are many pieces of commercial software and types of storage media (hard disk, magnetic tape, optical disk, solid state, remote storage) that can be used as backup facilities.

7.5.2.5 Managed Backup Services

One option that some organizations consider is the use of managed backup services. There are advantages and disadvantages with this approach:

- Advantages include that the organization uses less of its resources implementing and managing backups, less time in keeping up with the advances in technology and the maintenance of ICT for backups, the approach can reduce the organization's risk of data loss due to a lack of focus under operational pressures, human errors and/or accidents, it reduces the day-to-day responsibilities of backups and it provides a way of achieving the automatic availability of off-site for added protection.
- The main disadvantages are that the organization is entrusting its information assets and the management of the risk to a third party and not to internal staff or another part of its organization and the liabilities this entrustment might entail, as at the end of the day, the organization is still accountable for these information assets, which could involve information owned by its customers or suppliers and certainly could contain personal data.

If this is the option that the organization selects, then careful consideration needs to be given to the managed service contract and SLA to ensure the organization is adequately and suitably protected.

7.5.2.6 Backup Processes

In developing a best practice, a process for backing up information should consider the following questions:

- What are the organization's objectives for backup?
- Who is responsible for the backup process?
- When are backups carried out?
- How is the storage of the information to be organized and managed? What storage media should be used?
- What are the recovery objectives? For example, what is the desired time objective between the disaster occurring and the recovery and restoring of information in the system?
- What are the performance requirements?
- What are the arrangements for the backup process to tested to check the validity of the recovery of information?
- Do you keep backups secure and off-site from the main business premises?

- Do staff receive some training in doing backups to safeguard their essential information on their own computers that is not backed up centrally as well as, if necessary and appropriate, or knowledge of how to use the central backup system as per the backup procedures?
- Is there a need for backups to be encrypted and/or password protected to ensure their confidentiality, integrity and availability?
- Is a sample of the backup copies archived every 2 to 3 months?

Finally, always consider and plan for the worst-case scenario, when a disaster occurs, information is corrupted, damaged or lost. How easy is it recover the information?

7.5.3 Capacity Planning

In the context of information security management, capacity planning is the process of identifying the current and predicted future demands of system resources to ensure that the organization has sufficient capacity to do its business in a productive, efficient and effective way. The objective of capacity planning is to avoid the problems of overloads and reduction of performance levels due to the lack of capacity commensurate with the demands of users and customers. The capacity planning helps to minimize this discrepancy between capacity and demands.

An organization's information processing and/or networking demands will and do vary due to various reasons:

- Extending the processing operations of the business, for example, due to increase in sales and production, increase in the customer base, expanding the use of supply chains;
- Additions or modifications to the current internet services it offers and its outsourcing services;
- The introduction of new working practices and/or methods of processing such as allowing more staff to work from home and giving them more remote access;
- New computer systems being networked together or new networks being included;
- Redeployment of staff and a large increase in the staff count.

The term capacity, for the purposes of capacity planning, can be taken to mean, for example, the amount information processing that an organization needs to undertake in a given period of time, the amount of email traffic or the volume of internet transactions in which an organization needs to engage over a period of time.

Capacity planning is essential to ensure that the organization has sufficient capacity to meet the demands of business operations and to avoid problems such as lack of resource availability, system overloads, downtimes or degradation in system processing capability. Capacity planning is an important and critical management control in any risk management strategy and one that needs to be attended to and reviewed on a regular basis. The ISO/IEC 27001 best practice control A.8.6 addresses the issue of capacity management.

7.5.4 Change Management

Change management, for the purposes of this chapter, covers changes to those business processes, systems and services that support the processing of information in an operational environment. Our concern here is the impact that any change might have on information security in the operational environment. A change management process should be in place that identifies the change, assesses the potential impact of the change, ensures formal agreement to the change after reviewing the potential impact and takes the necessary actions to implement the agreed change. Adding to this, the change management process should take into account any information security concerns, requirements, risks and impacts related to the proposed change and verify that any implementation of the change meets these information security aspects.

There are many types of changes that the organization might need to consider either following an internal requirement or as a result of an external requirement. For example, changes regarding:

- Business mergers, downsizing or expansions and increases in staff levels;
- New business processes and deploying new ways and methods of working;
- Adoption of new technology, integrating new systems with legacy systems and deploying new network services;
- Third-party service delivery;
- Supply chains;
- Contractual obligations;

- New or revised legislation and regulation.

7.5.5 Third-Party Logistics and Supply Chain Management

Over several decades, the business of transportation and shipping, warehousing and storage and supply chain and inventory management through external service providers has been a growth industry throughout the world. These services have given a multiplicity of business benefits, allowing trade to flourish through the global movement of goods and products. Today's world is highly dependent on supply chains: many of the goods we buy as consumers rely on supply chains; the manufacturing of products relies on supply chains to deliver raw materials, parts and components for the finished product. Many parties can be involved in a supply chain: producers, logistics companies, distribution centres, retail companies and warehousing companies. The effective management of supply chains can reduce costs, help organizations remain competitive and make businesses efficient.

Protecting business information that is processed and flows through the chain is critical to its effectiveness and efficiency. Cyber-attacks on part of supply chain can have a detrimental effect on the whole chain. Therefore, understanding the information security requirements and the security management of the chain is important. ISO/IEC 27001 risk management is helpful in understanding the risks involved for each party in the chain and managing their mitigation.

7.5.6 Third-Party Services

7.5.6.1 Addressing Information Security

More organizations are using third-party services for various business functions and applications. The following are a few of these uses:

- Outsourcing call centres and help desks;
- Managed security services;
- Recruitment of staff taking over from the activity of a human resources group;
- Managed data backup and data recovery services;
- Hosting websites;
- Development of software;
- Recovery of debts and goods relating to unpaid invoices and payments;

- ICT support services.

Whatever the purpose of using external services is, it is important that the requirements for information security shall be addressed and be the subject of contractual agreements and SLAs. At the core of this is the information that is outsourced and processed by third parties and how this information is to be protected. The organization gives to the third party access of its information assets: information that might be commercially sensitive, operationally critical or personal data. This arrangement is an example of transfer of risk. However, this arrangement does not obviate the organization's ownership of the information or their responsibility to be accountable for the protection of this information. Having said this, there are many business benefits to using third-party services against which the organization needs to weigh the pros and cons as part of its risk decision-making process.

Considering information security, the organization should ensure that its information assets are given information security measures that are commensurate with the organization's business risks and to their business interest.

7.5.6.2 Service Agreements

The main aspects that should be considered with third-party suppliers in agreeing to a service agreement include:

- A clear definition of the security requirements and the security roles, responsibilities and obligations of both parties for fulfilling these requirements;
- The process for dealing with information security incidents;
- The access arrangements to ensure the confidentiality and accessibility of the organization's information assets;
- Backup and business continuity arrangements to ensure the integrity and availability of services and of the organization's information assets;
- Regular reporting;
- Provisions for management changes;
- Right to review and audit the third party's information security arrangements.

The organization needs to work with third-party suppliers towards a commercially amicable agreement of how to deal with the information security requirements. Clearly, the legal managers on both sides should be involved. The third-party supplier is more than likely to have standard contracts and SLAs in place, which may or may not include clauses for the provision of information security management. The organization and its legal department will need to review these third-party contracts and SLAs to check whether or not they comply with the organization's requirements.

7.6 Incident Management Narrative

7.6.1 Events That Compromise

Information security incidents can cause an organization's systems to be severely damaged, their work significantly interrupted, they can involve the organization in legal action (both civil and criminal) or, in the worst case, they might even threaten the organization's survival as a business.

The range of information security incidents is quite diverse and includes any event that could compromise the protection of the organization's information. This could be an incident caused by someone inside the organization (an insider threat) or someone outside, external to the organization. The incident might be caused intentionally or by accident. For example, an inexperienced employee using a badly written procedure, the intention of which is to give instructions on sending file attachments to email messages, might cause the incident. The incident might be a system failure or malfunction due to lack of proper maintenance, or a virus or other malware attack through lack of up-to-date anti-malware software. The incident might concern a poor access control system, a badly implemented procedure to prevent identification theft, a flawed process to prevent fraud or lack of physical protection allowing some unauthorized person to enter a secure area.

The severity of the incident could be minor or major, depending on the level of risk and impact involved. Incidents might occur daily, weekly, or monthly due to their uncertainty and so it is important that the organization has a process in place to handle the incident as and when it might occur. All incidents should be dealt with in a timely way, but there are some incidents that are more severe than others and so they should be given higher priority to resolve. The longer an organization avoids taking appropriate action, the worse the problem might become and the greater the resulting cost; timeliness is critical to protect the organization.

7.6.2 Use Cases

7.6.2.1 Importance of Supply Chain Risk Assessment

A medium-sized auto-engineering supply company was experiencing a number of external unauthorized attempts to its internal networks. None of these were successful, but management was worried by this concentrated spate of attacks. Their firewall detecting then followed this and blocked a rapid increase in the number of malware-based attempts to penetrate their system. At the same time, their email system was being bombarded with a continuous stream of thousands of unwanted emails, enough to overload their system. This chain of events further worried management, as it seemed that their company was being targeted. Fortunately for the company, their incident handling process was able to deter any serious impacts.

It turned out, after consulting with one of their end customers, that one of the customers in particular was the main target and all the customer's suppliers were indirectly targeted to interrupt the supply chain. The incident involved the customer's supply management system being compromised. After many weeks of investigation involving the end customer and three of its suppliers, several men were charged with criminal offences, including employees from all the companies involved. The impact on all companies involved was costly and resource-intensive. The media coverage reporting also had an impact on the company and its suppliers.

Finally, over a period of time, the situation got back to near normal and the company started building back the confidence and loyalty of its customers. All those involved learnt many lessons in handling this supply chain incident and this should help them in avoiding future incidents such as this.

7.6.2.2 Importance of Testing

A medium-sized company producing a range of electronic components was hit by a spate of system problems. Their design and production lines were highly dependent on the use of computer-aided design (CAD)/computer-aided manufacturing (CAM) applications for the design work and their automated assembly and testing facilities. The processes seemed to become erratic and unreliable and produced faulty results, which slowed down productivity and, on two occasions, the complete system had to be restarted.

The company's incident team was assembled to resolve the problem. In collecting all relevant information about these incidents, they discovered the problem seemed to be related to a few new customized application software packages that had been deployed 2 months ago but had never been tested before being used in an operational environment. The supplier of this software was immediately contacted and asked to solve this problem. Over the next 24 hours, software engineers found a number of bugs in the software products and produced several patches to resolve the issue. The customer tested these and found them now to be functioning as intended.

The company got back to its normal operations and productivity after 3 days of downtime. Since then, several improvements were made to their information security and software management controls.

7.6.2.3 Importance of Conflict Resolution

The Online Services Company recently introduced a strict Internet Usage Policy to improve staff productivity and customer services and to direct staff to spend less time surfing the internet for personal use during busy times in the office. This seemed to work for a short period of time and then a small group of staff became disgruntled and at odds with management. Shortly afterwards, the company seemed to experience an increase in the number of unauthorized access attempts during the first two weekends in March. There was also an increase in the number of internally sourced blocked viruses.

On examining the internet usage records, management noticed a pattern between blocking external access attempts and internal access attempts. More detailed analysis of other monitoring records kept by the organization showed that those attempting internal access were the same people who expressed their strong objection to the recently introduced policy and three of these people worked during the weekend of March 3 and 4.

The management interviewed the staff and then started disciplinary procedures. During the process, the names of friends who were involved in the external attempts finally emerged. It also emerged that some of the unexplained loss and damage of company files could be traced back to the activities that took place over the weekend of March 3 and 4.

The relevant authorities were called in to help in the case and to make charges where appropriate. This group of collaborators, staff and friends was responsible for gaining unauthorized access, destroying files and planting viruses in the company's system. The enquiry concluded with internal staff and external individuals being charged by the police for criminal activities

7.6.3 Processes

The general process of information security incident management is illustrated in Figure 7.1.

7.6.3.1 Planning

Being prepared is vital to being effective in the event of a major incident. This is a fairly commonsense thing to say, but taking the right action to being prepared is not often done in practice. In many cases, producing a set

Figure 7.1 Incident management process.

of procedures for incident handling and having them tested and tuned to the organization's needs more often or not take place after a few major incidents have happened. Some organizations partially address the problem; for example, they may have an incident system in place but nothing else to protect against malware or software vulnerabilities and no comprehensive system that can deal with any form of incident.

So it is important to have in place a set of policies and procedures defining a typical set of incidents, the process for identifying and evaluating these, what the response, resolution and recovery process is and a what's next process, who is responsible for what part of the process and the means of communications during an incident occurring. It is important that this is not just a paper exercise, but these processes are tested in advance of any potential incident occurring.

The planning activity should establish an incident response team with defined roles and responsibilities. This planning activity should also define who to engage with regarding incident events: internal groups (senior management, security team, IT support, legal team, media relations, human resources) and external groups (affected external parties, software vendors, computer emergency response teams (CERTs), internet service providers (ISPs)).

7.6.3.2 Identify, Detect and Report

The process needs to identify, detect, report and record:

- What the incident is (is it minor or major)?
- Where in the system is it occurring (internally, externally or both)?
- What resources are being affected (internal and/or external)?

There may be a precursor, a sign, that an incident may occur in the future, or an indication, a sign that an incident may have occurred or may be occurring now. Accurate reporting and recording of this information on an incident reporting template are essential to the analysis and responding phase of the process.

7.6.3.3 Analysis and Evaluation

The information that is reported needs to analysed and evaluated by the team to assess the extent and scale of the incident and the severity of the incident and to ask questions such as:

- Is what is being reported an actual information security incident or some other type of event?
- Is the information accurate? Does the information reported provide a complete picture of the situation? Some indicators may provide information that is incorrect or inaccurate, incomplete, misleading and ambiguous and introduce false-positives or false-negatives.
- What is the nature of the incident? What are its characteristics? Are the symptoms clear, obvious or not easy to understand?
- What is the scope of the incident? The scope should indicate which parts of the business are affected, what information systems and business processes are affected, what networks, IT systems, applications, services are affected and the who, what and how (i.e., source and cause of the incident and how it is occurring).
- Is it contained in one area or is it spreading?
- How fast is it spreading?
- Is it isolated to just internal systems, has it come in from the outside or is it spreading to outside?

These and many other questions need to be asked to be able to start the response and recovery process knowing what should be done, where and to what. Also, both internal and external stakeholders need to be notified.

The incident response team needs to analyse, check and validate each incident. On deciding whether the event is actually an incident, the team should do initial analysis to determine the scope of the incident and sufficient information to prioritize what actions need to be taken and what response strategy to take (e.g., containment, eradication).

7.6.3.4 Response

In responding to the incident, the team needs to work in a timely way and to recover the system to its original operational state as soon as possible. Time is of the essence to stop the problem spreading and to ensure the organization can keep functioning properly in order not to damage or lose its business through excessive delays caused by the response process.

Some key points are:

- Prioritization of what to deal with first to respond to the incident is of crucial importance; expedient management decision-making is critical to the success of the response, resolution and recovery; the incidents to be dealt with as high priority will depend on the business impact and recoverability requirements, operational impact, information security impact, corporate and financial and legal impact (recoverability), but people safety and security should get the highest priority, above all else.
- The nature, severity and spread of the incident and the resources affected determine the time and effort needed to get the business operations, processes, systems and networks back to normal. The time and resources to recover may be extensive.
- Get the information security handling team started on dealing with recovery.
- Process the evaluated results and evidence about incident.
- Take action to contain the incident.

7.6.3.5 Timeliness of Response

An incident can be thought of as the coming together of a threat and a vulnerability. When a hacker is able to exploit the weaknesses in an organization's access system software or an access method, the result in an incident of unauthorized access. The impact can vary depending on what the hacker is able to do and how fast the organization is at stopping the hacker, in the case of malware, how fast the organization is able to stop the spread of the malware infecting IT systems. Therefore, time is of the absolute essence in dealing with incidents, especially those that are severe and involve critical business systems or processes.

The hacker may start by gaining access to files and downloading information, and then he might install malicious code in the system and finally use the organization's network vulnerabilities to connect the network of one of its major customers. This whole chain of events, if allowed to happen, could result in various financial and legal impacts as well as loss of its image and reputation. If this chain of events is contained early enough in its cycle, then the damage and impact to the organization can be limited. Therefore, the timeliness of response is very important.

If the incident puts people's lives at risk, their safety and security are the highest priority, but, again, timely response is of the essence. Such might be the case with an environmental incident, an explosion, a bombing or a natural disaster. The people aspect takes highest priority over everything else.

7.6.3.6 Recovery

Recovering from the incident involves getting the system that was affected back to its normal operational state so that business can continue to operate. Again, time is of the essence to avoid any unnecessary losses to the organization during the time when the incident was active and the organization was not fully operational. Some key points:

- Contain, eradicate and resolve the problem.
- Reestablish any affected computer systems and other IT equipment, the organization's computer networking capability and connections; recovery may involve:
    - Restoring from backups and restoring any lost, damaged or corrupted data that occurred during the incident;
    - Rebuilding systems;
    - Replacing files that have been compromised;
    - Installing patches;
    - Changing passwords and accounts and improving system and network security;
    - Reconfiguring systems and networking.
- Reestablish and restore business systems, processes and services back to a normal operational state; check that everything is functioning correctly and as expected; communicate this to staff and management, customers, suppliers and other appropriate stakeholders.
- Fully document the incident case and changes that need to be made to the system, policies, procedures and processes; undertake any final actions to close the incident case file.

7.6.3.7 Post-Incident Review and Follow-Up

This part of the process is important first to do a review of the incident situation and second to take preventative actions to make the necessary improvements to protect the organization from further incidents of this type from reoccurring.

Post-incident review:

- Review the incident report forms.

- Carry out a forensic analysis of the incident.
- Review the effectiveness of the incident handling process and the information security incident handling controls.
- Undertake a business impact analysis of the incident.

Improvements and updates are to:

- Make recommendations for improvements to the ISMS.
- Take steps to improve the incident management process and controls.
- Update policies, procedures, processes and training where appropriate.
- Where necessary, make changes to contracts and SLAs.
- Finally produce an incident report.

7.6.4 Incident Investigations

Incident investigations involve the use of digital investigation techniques, to identify, gather, evaluate and preserve information (digital evidence: information stored or transmitted in digital form that may be relied on as evidence) related to an incident that has occurred. Such information might be used as evidence, for example, in the case of civil or criminal actions, or for internal management purposes (e.g., internal policy breaches and violations, resolving operational problems and for system recovery).

Examining and investigating are mentioned in ISO/IEC 27001 Annex A.5.28 "collection of evidence" and are covered in more detail in the standards referred to in Section 7.6.7.

7.6.5 Incident: Cause and Effects

The incident is the situation in which an organization needs to take action today based on what happened in the past and its current experiences of the present; yesterday's causes become the results of today (the effects) and tomorrow's results are caused by today's events. The causes and effects are both events: the cause of the security compromise yesterday has an effect on the organization today; the effect of not resolving this problem today causes a continuing problem or a set of new problems tomorrow. We should learn from the mistakes of the past that cause our security problems to help us do the right things today and to protect our future; see Figure 7.2.

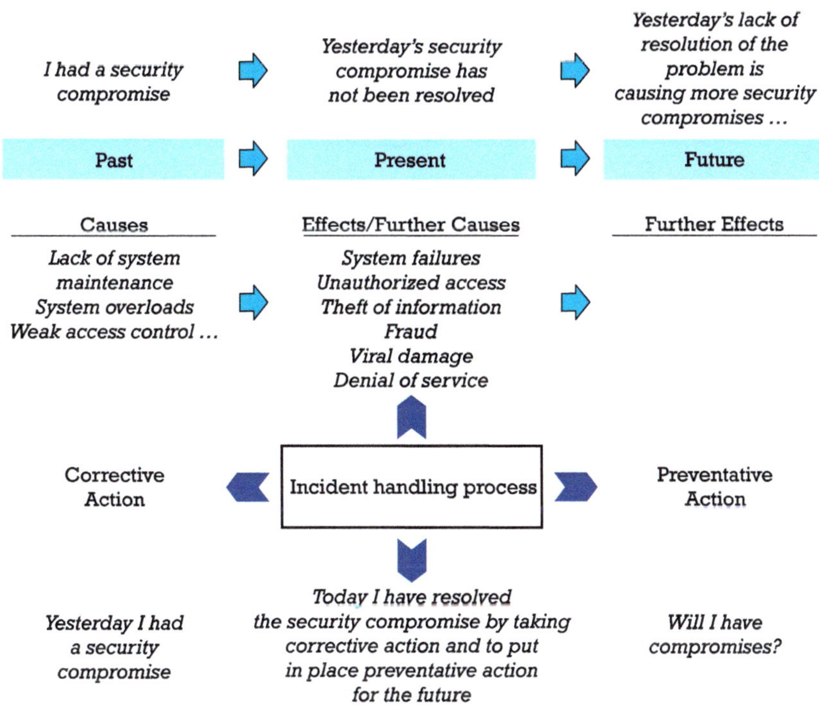

Figure 7.2 Incident: the cause and effect.

7.6.6 Incident Management Team

An information security incident management process involves those activities to identify, analyse and correct problems and facilitate the response and recovery of the systems that have suffered from the incident. This process should involve a multidisciplinary team, with a range of skills needed to deal with a range of issues. Typically, the incident might require the involvement of management, information security staff, IT and non-IT staff, physical security staff, human resources, legal advisors, those handling external relations and the press and sometimes the external party's emergency services, policy and maybe even trade unions.

The team needs to be able to analyse the situation, determine the scale and severity of the compromise and initiate appropriate action to correct the problem and prevent it from reoccurring and to liaise and communicate all the relevant internal and external parties. The incident handling team should have a leader whose task it is to coordinate the process and to facilitate and manage the team that is deployed in the process.

7.6.7 Standards

More detailed technical information on information security incident management can be found in the following.

7.6.7.1 ISO/IEC

- ISO/IEC 27035 Information Security Incident Management:
  - Part 1 Principles of Incident Management;
  - Part 2 Guidelines to Plan and Prepare for Incident Response;
  - Part 3 Guidelines for Incident Response Operations.
- ISO/IEC 27037 Guidelines for the Identification, Collection, Acquisition and Preservation of Digital Evidence;
- ISO/IEC 27038 Specification for Digital Redaction;
- ISO/IEC 27041 Guidance on Assuring Suitability and Adequacy of Incident Investigative Methods;
- ISO/IEC 27042 Guidelines for the Analysis and Interpretation of Digital Evidence;
- ISO/IEC 27043 Incident Investigation Principles and Processes;
- ISO/IEC 27050 Series Electronic Discovery.

7.6.7.2 NIST

- SP 800-61 Computer Security Incident Handling Guide;
- SP 800-83 Guide to Malware Incident Prevention and Handling;
- SP 800-86 Guide to Applying Forensic Techniques to Incident Response.

Note that this list of standards and guidelines is not exhaustive, and there are many others available from different sources, including standards bodies, industry associations, user forums and computer emergency response organizations.

7.7 ISMS Availability and Business Continuity Narrative

7.7.1 Value and Importance

There should be little doubt as to the importance and value of ISMS availability, information security incident management and business continuity play in supporting the organization.

We have many global examples of business systems, processes and services failing and being inoperable for hours, crippling businesses whose operations come to a halt. The causes of these incidents can range from natural disasters to online attacks with the deliberate intent of bringing systems down. Today's business reliance on the internet and IT should place more of an emphasis on businesses being ready to respond to such incidents and to recover as soon as possible to ensure an acceptable level of continuity of business operations. The business impact of not having appropriate resilience, response and contingency and protection processes in place can all too well be seen by the various real cases that have hit the world headlines from human-created situations to natural hazards or a combination of both, for example, where a natural hazard triggers a disaster relating to some human-created situation. Headline news raises awareness of how vulnerable our systems can be and the impact on business when things go wrong. This heightened awareness draws the attention and concerns of organizations, the citizens and consumers alike. It triggers some organizations to rethink how they would cope if it happened to them. They need to question and examine how good their processes and systems would be at coping with these issues.

There are many companies that have not sufficiently protected themselves with an effective business continuity process and some do not even have such a process in place. Sometimes the reason why the business continuity is not in place is that some think that it presents too many challenges and is too expensive or they do not know what they need to do. These are particularly some of the reasons why small-sized companies (SMEs) do not have business continuity because the cost, know-how and resources are out of their reach. This should not be the case because there are many routes available for getting simple advice on what can be done and the implementation of some of the basics should not be out of the reach of SMEs. At the end of the day, the company need to weigh the costs of the resulting impact of not doing anything against the cost of doing something and its reduction in the business impact. It could make the difference between survival and the closure of the business. It should be clear that there is a business value in doing it no matter how perfect the solution is. It is important to take that first important step of doing something to build upon the basics, and the rest will fall in place with the right guidance. One of the important things to do first is to get management commitment and this can be achieved by doing an initial business impact analysis.

7.7.2 Business Impact

One important question to ask is: What are the critical things that are essential to the organization to enable it to continue to operate? Once we know this, we can then test these by looking at some examples that might be appropriate to the organization running its business. These can then be used to calculate the business impacts with which the organization might be challenged in the face of adversity in the event of disastrous incidents occurring.

If the organization is not prepared and has no continuity plans in place for the advent of a disaster, disruption to its operations, a major system failures or external influences from the markets in which it is involved, then it leaves the organization wide open to being exposed to high impacts which could ruin the business. For example:

- Supply chain failures and disruptions;
- Denial-of-service attack shutting down or grossly restricting operations dependent on ICT systems, services and a networking capability;
- Critical loss of supporting infrastructure (e.g., electricity, gas or water supplies) and services (e.g., telecommunication services, ISP services, outsourcing facilities);
- Major economic failures and losses, including critical depression in market conditions (e.g., share prices, interest rates, world/country economies);
- Strikes, political activists, breakdown in workforce relations;
- Environmental hazards and disasters;
- Bombs, explosions and terrorist attacks;
- Major legal, regulatory or contractual failures leading to penalties and liabilities.

These examples cover financial, operational, customer/supplier, staff-related and legal issues.

7.7.3 Plans

It is essential that the organization should have in place a business continuity plan. This plan should include documented procedures that will give clear instructions of what to do to respond, recover, resume and restore to an acceptable level of operations following a disruption, disaster, emergency or failure. This plan should include information on the required resources and services, and activities to be undertaken, to ensure the continued use of critical business functions, processes and systems.

The scope of the business continuity plan is likely to cover a broad range of topics and activities, especially provide availability, continuity and recovery of critical business elements: business functions, processes and services. The scope of the plan should cover aspects of disaster recovery, incident management, information security management, safety management and emergency and crisis management.

7.7.4 Processes

Business continuity planning and supporting processes are aimed to ensure that the right resources are available and the right activities are being undertaken to keep continuity of the critical elements of the business running and operating, and if these critical elements become unavailable and inoperable during a disaster, those processes are in place to recover these as a matter of urgency.

In today's business environment where IT is a vital tool to the proper functioning of the business, it is important to protect the IT accordingly in the event of a disaster. When the IT has failed or been compromised or damaged or has become unavailable during a disaster, then timely recovery of the IT is vital, which might mean replacing damaged IT.

An essential process included in business continuity management is undertaking an impact analysis. This involves gathering, analysing and evaluating this information to assess the impact with regard to loss of confidentiality, integrity and availability. The results of this can then go towards a review and decision-making process to determine the organization's requirements and what should be done to guard against such business impacts occurring.

For more detailed information, the reader is directed to ISO 22301 and ISO 22320 (see Section 7.7.5).

7.7.5 Standards

More detailed technical information on business continuity and information security incident management can be found in the following.

ISO:

- ISO 22301 Business Continuity Management System—Requirements;
- ISO 22313 Business Continuity Management Systems—Guidance;
- ISO 22320 Emergency Management—Requirements for Incident Response.

ISO/IEC:

- ISO/IEC 27031 Guidelines for ICT Readiness for Business Continuity;
- ISO/IEC 27035 Information Security Incident Management.

NIST:

- SP 800-34Rev1 Contingency Planning Guide for Federal Information Systems.

7.8 ISMS Use Examples

7.8.1 SME Design Services

7.8.1.1 Introduction

A small company with 11 full-time staff provides household designs for domestic users. This includes custom-fit kitchens and bathrooms, studies, living rooms and dining rooms. The company consists of consultants and designers, salesmen, showroom staff, the owner and a secretary.

The company has been in operation for 7 years and is well-recognized as a supplier of quality designs. The company has just one showroom/office located on the ground floor of a 5-floor building on a busy commercial road of a town in the United Kingdom. The company has already operated an ISMS for the last 3 years. They now want to expand the business to network its PCs, give remote access via consultants' laptops when on customers' sites, add website services to its customers and open two other offices in other European countries to provide the same sort of services.

The company has carried out a strengths, weaknesses, opportunities and threats (SWOT) analysis to assess the feasibility of the objectives and potential success of expansion. Based on this analysis, it decides to go ahead. Their business implementation plan includes an extension of the scope of its ISMS to include its website presence, remote access and its two new planned offices.

7.8.1.2 Existing and Future ISMS Deployment

They went through a gap analysis to identify what their existing information security system includes:

- An information security policy;

- Procedures on backups and viruses and handling personal and customer information;
- Password control on all office PCs and consultant laptops;
- Physical access controls to its offices, rooms and filing cabinets.

A target ISMS scope is defined to cover all existing operational needs as well as their future plans for expansion. A risk assessment on the new ISMS scope taking account the increase in its resources and the expansion of its business processes, the results of which are evaluated and developed into a risk treatment and ISMS implementation plan. As their future business plan rolls out over the coming 6 months, so does the implementation of their ISMS and, within a further 6 months, they are fully operational. Their internet side of the business starts to grow as they are getting more customer enquiries and orders, as does the number of staff employed by the company.

7.8.2 Legal Services

7.8.2.1 Introduction

A small legal firm with 17 full-time staff provides legal advice for both commercial businesses and citizens. The firm has three offices in different cities in the United Kingdom. It has been established for over 25 years and has many high-profile corporate clients on its books. Over the last 3 years, it has relied more on the use of IT services and the internet to conduct its business. It is now becoming quite clear that it needs to do something about enhancing its existing information security to take into account its high-tech way of doing business.

7.8.2.2 Existing and Future ISMS Deployment

The firm already has security measures in place for its business but these are stand-alone solutions with very little networking deployed. The security it has in place needs to be enhanced to meet the needs of management for a more integrated, networked environment. The decision to go down the route of using ISO/IEC 27001 was made after the firm was made aware of other firms in the legal services business going down the same route. Its main IT resources and internet capability consists of:

- A number of office PCs networked together to a central server;
- Internet access is via a firewall connected to the office server;
- Laptops, tablets and mobiles (all Wi-Fi enabled) used by legal professionals.

The information it processes includes:

- Client confidential case files, reports, letters, statements and other documents;
- The firm's internal confidential and restricted information, documents and files;
- Library of legal books, works of reference and generally other publicly available information.

The task on which they embarked (with the help of external consultants) was the first to identify the needs and expectations of management, clients and other interested parties relevant to an ISMS, then define an appropriate ISMS scope commensurate with the nature of their business and then carry out a risk assessment and risk treatment exercise. The risk assessment and risk treatment exercise concluded with the need to implement a range of new controls to enhance their existing protection and to facilitate protection of their new networked business environment. This led the firm to implement an ISMS to the requirements of ISO/IEC 27001 and for an audit to be carried out by an accredited certification body, which resulted in the award of an ISO/IEC 27001 certificate.

7.8.3 Electronic Accounting System

7.8.3.1 Introduction

The management of a large retail company decides to install a fully integrated electronic accounting system, but before doing so, the decision is made to check the requirements for achieving an effective level of information security. The current accounting system does use PCs, but these are not networked together to provide a streamlined start to finish process for handling sales, purchases, orders, invoices, payments, delivery and dispatch notes. Their future plans require the installation of an IT system networked throughout the company with links to external customers and suppliers, with appropriate software application packages. This will enable staff to process and share information online within the company and also to electronically process orders, payments and other documents to suppliers and customers.

7.8.3.2 Existing and Future ISMS Deployment

Although the company has protection measures in place, it decides to carry out a complete overhaul of its information security taking into account the existing controls. It recruits someone as its security officer and, together with departments such as sales, purchasing, finance, operations and IT services, it develops an ISMS scope and design based on ISO/IEC 27001.

It gets management approval and commitment to implement this ISMS design across all the five locations in which it has its business. As the project is quite a sizable undertaking, it wants to achieve this electronic accounting system development as soon as possible to take advantage of the business benefits of such a system and provide added value to its customers that it decides to supplement its internal project team with external consultants.

The project is not without its setbacks as commercial pressures indicating a greater need to have the system fully operational as soon as possible. In addition, the gap analysis and risk assessment was not as easy to accomplish due to multisite nature of the project, the current lack of coordination, consistency and integration of the current system and some local resistance from staff regarding issues with change and questions on security. It is then agreed that senior management must get more closely involved to show their full support and to commit to more resources. After several months' delay in making progress, the company rolls out a fully operational electronic accounting system with information security in place. Due to this experience, management is persuaded to recruit a second full-time employee to support the security officer to maintain the ISMS. The company over the next 6 to 9 months prepares itself for third-party certification, which it achieved.

7.8.4 Government Payment System

7.8.4.1 Introduction

An e-government department in Southeast Asia operates both off-line and online payment systems refunding citizens for their Social Security claims. As with many governments around the world, the attraction is converting its business processes to utilize e-commerce technology and providing the public with online access to public information and government services, for example, to pay taxes and claim benefits and apply for car licences and visa applications. Such use of e-commerce can be used for internal government business, government to external organization business and transactions or government to citizen transactions and sharing of information. In this scenario, another government department whose responsibility it is to advise on information security matters recommends that all e-commerce applications go down the ISO/IEC 27001 ISMS route.

7.8.4.2 Existing and Future ISMS Deployment

This particular government department, dealing with Social Security payments, always had information security measures due to sensitive nature of the claims it was handling, especially as it involved general personal data and more specific details such as health-related information. Its future plans had covered moving over to a more electronic way of submitting claims with a Social Services Internet Café-type facilities in government centres and offices around the country all networked to the department dealing with social services. The ISMS for this Social Services Internet Café had to deal with citizen privacy technology, citizen smart cards for identification and authentication purposes as well as accessing the system, secure exchange of citizen information and secure information processing facilities in the concerned government department. Around this is the management framework for making this function effectively.

The scale of this project is large and complex, so it is decided to roll this out in phases focusing on one particular town of a medium-sized population to trial the ISMS design to improve the design where necessary and then to proceed with other towns to gradually cover the whole country. This first trial was successful, although the ISMS did need various adjustments and improvements regarding its implementation controls. After subsequent phases, work had been completed, growing the ISMS solution to over 70% coverage on the population, and the government decided to have a third-party certification audit.

7.8.5 Outsourcing Call Centre Operations

7.8.5.1 Introduction

In the interests of streamlining the business and cost-cutting, a major international European-based company outsources its call centre operations to a company in Asia. This type of business has become very common and popular over the years and one that is not expected to change but to grow. The outsourced call centre operations cover the company's customers all over Europe and the daily number of customer enquiries continues to grow due to their business successes.

7.8.5.2 Existing and Future ISMS Deployment

It is of the utmost importance that an information security risk assessment is carried out before undertaking such an operational change, especially considering the various factors involved in outsourcing to overseas countries, including the following.

Geographical, environmental and political factors:

- What hazards apply to this particular area of the world? Is it in an earthquake zone or an area prone to volcanic activity, tornadoes, typhoons, tsunamis and other natural phenomena? Many areas in Asia have these types of problems.

- Is the location in an area of political and social stability? Is it economically stable?

IT facilities, communications and levels of service:

- Are the communication systems able to cope with the volume of customer traffic?
- Can the outsourced company supply adequate facilities and resources to fulfil the contract requirements (e.g., sufficient capacity to meet demands, effective information security, backup, disaster recovery and business continuity arrangements, response times, performance and service levels)?
- Can the location and its facilities be easily secured from a physical perspective from local threats?

Information security:

- Has the outsourced company carried out a risk assessment and risk treatment exercise? Has the outsourced company implemented adequate information security to protect the information of its customers? Does it provide appropriate information separation for each of its customers? Does it take regular audits and reviews of its information security? Does it have an incident management process in place? Does it have a business continuity management system?

Staffing levels:

- Can the outsourced company supply an adequate number of trained staff to fulfil the contract requirements (e.g., communication skills, knowledge, experience and language skills)?

In the case of this particular outsourcing organization and its information risks, it has developed an ISMS, it is being used to provide protection of its call centre operation and it has been certified to ISO/IEC 27001 to demonstrate to its customers that it is fit for purpose. Sometimes the customer itself requires the outsourced company to be certified to ISO/IEC 27001.

7.8.6 Manufacturing Systems

7.8.6.1 Introduction

A successful medium-sized manufacturing business is certified to ISO 9001 (Quality Management System (QMS). It has now decided that it should consider applying ISO/IEC 27001 across its business to complement its ISO 9001 capability. It would eventually like to integrate the two systems together to create an efficient and effective management framework to gain the cost benefits of having an integrated certification audit.

7.8.6.2 Existing and Future ISMS Deployment

Its current information security practice seemed to have been sufficient in previous years, but there is growing awareness of the threats and risks it faces from observing market trends in its particular industry sector. An additional drive towards gaining more awareness was the fact that one of its competitors suffered a damaging attack on one of the competitor's IT-based automated production lines. It employed a consultant to do a gap analysis and risk assessment of its current systems. The results of this were quite startling for the company; its current information security provisions were only 35% compliant with ISO/IEC 27001 and 20% of these had been picked up by the recent ISO 9001 surveillance audit as things that had an impact on its QMS implementation. Even more startling were the large areas of noncompliance, 55% of which turned out to be high-risk areas based on the results of the consultant's risk report. The main areas of risk were:

- General lack of management policy and procedures; no operational procedures in a number of key areas such as for information handling, backups or access control;
- No regular risk assessment programme;
- No regular monitoring, audit or reviews of its information security;
- Insufficient access control for operating systems, internal networked services and applications;
- No information security incident management process;
- The gateway to the internet being wide open to external threats, allowing relatively easy access to its internal network;
- A lack of information security awareness and training and insufficient resources to carry out information security-related tasks;
- A lack of information security considerations in their contracts and SLAs with third-party service providers;

- No vulnerability management;
- A lack of protection against social engineering attacks;
- A lack of information security for business continuity.

After several management decision-making meetings, the business decided to go ahead with developing an ISMS for the whole company. It contracted the work to a small company specializing in this work, which had a tracked record in different parts of the world in helping clients to establish an ISO/IEC 27001-compliant system. After an intensive contracting period, the business had implemented an ISMS and became certified to ISO/IEC 27001 before its competitor. Its next challenge was to work with its certification body towards planning an integrated ISO 9001 and ISO/IEC 27001 audit.

7.8.7 Supply Chain Management

7.8.7.1 Introduction

Supply chains are the norm of most organizations. The organization needs to work efficiently; it needs to manage its operational and business relationship with its supply chain effectively, especially if its business is in a fast-moving and dynamic market in which market competition, customer satisfaction, timeliness and profitability are key. The supply chain can be viewed as a networked system of organizations, human resources, operations and communications, information and IT resources all working together in delivering a product or service to the end customer. The supply chain industry thrives on the flow of accurate information being available 24 × 7 (that is, 24 hours a day, 7 days a week).

Typical examples are online purchases of books, food supply chains, car parts, components and subassemblies to support car manufacturers, energy supplies, supply chains supporting the construction industry, supply chains supporting the travel industry such as for airline travel or chains supporting the production and supply of medicines for the healthcare industry. All of these supply chains need to run both efficiently and effectively to fulfil customer and consumer needs and demands.

Supply chains link together an organization's value chain of activities such as the logistics of both its inbound and outbound products, its operations in producing things or delivering services, its marketing and sales support and its maintenance support.

Outsourcing companies (sometimes referred to as third-party logistics provider (3PL)) are available to support organizations' supply chains such as in the provisioning of integrated warehousing and transportation services.

7.8.7.2 Existing and Future ISMS Deployment

Again, supply chain management is another application in which ISO/IEC 27001 is able to provide a management framework to manage its information security risks. Some organizations that supply logistic support services have been certified against ISO/IEC 27001 for this very purpose, especially those that also provide international logistic support.

CHAPTER

8

Contents

8.1 Performance, Change and Improvement

8.2 Monitoring and Operational Reviews

8.3 ISMS Measurements Programme

8.4 Ongoing Risk Management

8.5 ISMS Internal Audits

8.6 Management Reviews of the ISMS

8.7 Awareness and Communications

Performance Evaluation

If you cannot measure it, you cannot improve it.
—Accredited to Lord Kelvin, also known as William Thompson (1824–1907)

8.1 Performance, Change and Improvement

8.1.1 How Effective, Adequate and Suitable Is the ISMS?

Is my ISMS fit for purpose? What exactly does this mean? Here we are taking it to mean that the performance of the ISMS is effective, adequate and suitable to meet with the organization's policy, strategy, objectives and business and legal requirements. This leads to the question of how I evaluate the ISMS to check its performance, and the answer is by monitoring aspects of the ISMS, taking ISMS measurements and carrying out audits and reviews. So we might ask: Are the ISMS processes and controls successfully providing protection of the organization's information resources? Are the policies and procedures suitable? Are the ISMS resources adequate to manage the risks? These and many other questions arise when we start to evaluate the performance of the ISMS to check that it delivers the desired and intended results and outcomes of the ISMS. Does the ISMS implementation address all the requirements and issues discussed and managing the risks as discussed in Chapter 4? Has the organization implemented ISMS with suitable and adequate policies, processes, procedures and controls to manage the risks discussed in Chapter 4?

Performance levels of the ISMS will vary as time goes by because of change: business changes, market changes, political and economic changes, resources changes, environmental changes, legal changes and technological changes. These are a few of the types of changes that the organization may face, and all of these carry with it the chance that the levels of risk and impact will change.

8.1.2 Change and the Certainty of Change

"Nothing is as constant as change," attributed to Heraclitus, an ancient Greek philosopher, is an old adage, which is very apt and simply means nothing stays the same forever. It is often said that change is the quality of impermanence and everything on this Earth is impermanent, in a state of constant flux. The Greek philosopher Heraclitus saw change as ever-present, ever-flowing and all-encompassing. Change is at the core of continual improvement and making sure the organization is managing its risk to ensure it is protecting its information at all times. Monitoring and reviewing changes and the causes and effects enable the organization to manage the resulting changes in risks and impacts appropriately. The risk we assess at one point in time may change some short time later or a week or month later, and so the protection we have in place to mitigate the risk may no longer be adequate.

Changes may alter or modify a situation or may convert something into something else, and changes may happen slowly or rapidly. Change can be sequential, nonsequential or consequential. Organizations need to view change as a natural process and they should monitor and review the patterns of change and any other aspects that will affect and impact the organization.

As Sun Tzu's words reflect, the organization should know itself and its business and know the environment and markets in which they are doing business; this can lead to a better chance of achieving 100% success, and continuation of this success requires the organization knowing, understanding and adapting in a suitable way to both the internal and external changes. The organization knowing its information security strengths and weaknesses (vulnerabilities) and the threats that might exploit these gives it a better chance of managing its information security risks. Regular monitoring and reviewing any changes to these strengths, weaknesses and threats enable it to be in a better position to continue to improve and maintain an effective information security capability.

An organization knowing its online business capability and its information security strengths and weaknesses and its competitor's capability, strengths and weaknesses in this field enables the organization to manage and improve its information security risk controls' capability. Again, regular monitoring and reviewing of any changes to these conditions to improve and maintain an effective information security capability and enable business opportunities.

How much change has occurred? What are the effects of the change? What is the level of change in risk and impact? These and other related questions all become the subject of measurements as part of the overall performance evaluation exercise.

8.1.3 Change Management

8.1.3.1 Change Happens

There are two basic forms of change: change that is planned by the organization (an organization plans to adopt a new technology), and change which is unintended or unplanned by the organization and not under their control (changes in the market and legislation to which the organization needs to respond and adapt). In both cases, it is critical that information security adapts and/or responds appropriately to the change.

Change management is about the path of transition from what is the current state of the business operations now to the future state of business operations. The process of transition can go relatively smoothly, or it might be fraught with difficulties. As such, change management involves members of staff, business and operational teams, management and the organization itself. How an organization adapts to the change process can make a big difference to its business. If there is resistance to the changes, then this can hamper the overall success of the transition and the end result. Individuals, in general, are normally anxious about change and have an innate sense to resist. In the case of major business change (e.g., downsizing, merger, more automated business processes), the organization needs to help its staff to prepare and make the transition as smooth as possible whilst helping to allay any anxieties.

There will always be individuals or certain business groups, not the whole organization itself, that resist change: this situation is normal. Taking the path of least resistance is a good strategy for many difficult situations in all walks of life. Using an area of the business it knows to be open and receptive and which offers the least resistance to a proposed change is a good starting point to introduce and test the change is more likely to succeed. Once this area of the business gains confidence through deployment of the change and other parts of organization see its success and value to the business, the organization can then start to roll out the change across the rest of the organization. Amidst all this change, information security should continue to be effective.

It has been said that the key to success is to keep change to a minimum and not go for change for the sake of change. Some things may not or should not need to be changed, in order to ensure consistency and business continuity. Adoption of changes should be in flow with the business and going for a Big Bang approach is not necessarily the best move, rather change strategy via a path of least resistance. It is entirely up to the organization and its risk appetite and it is its commercial choice how much change is necessary to meet the future mission and business objectives. Whatever set of changes are proposed by the organization, an information security risk and impact assessment needs to be carried out before any changes are implemented. For example, a complete overhaul and upgrading of the organization's IT strategy for both IT acquisition and deployment needs a good knowledge of the internal and external factors which might have an impact on this strategy. In addition, the information security controls required in relation to the use of this new or upgraded IT needs careful consideration, such as knowing what external changes have driven this new strategy, what the changes will mean to its staff as well as to its customers and suppliers or what impact these changes will have on its market positioning.

8.1.3.2 Process

Change management process must consider:

- How the changes fit into the bigger business environment within which the organization operates, making sure the implementation of the changes work for the organization.
- There are no conflicts with the ISMS implementation, and all the necessary risks and impacts have been taken into account.
- There is adequate awareness and training of staff with regard to the changes, including training related to information security.
- Communication of changes to all interested parties.
- Gaining the support and co-operation of staff for the adoption of the changes.

8.1.3.3 Types of Change

There are changes driven by external forces to which the organization needs to decide how to respond and react, and there are changes planned and instigated by the organization, for example, where they can be more proactive in driving and leading the market with new customers, services and/or products. In both cases, business risks need to be considered, for example, the risks of adoption by the organization of a new technology to give it more advantage in the market or taking risks to drive the market with its own development of new technologies. Other examples include:

- Changes in business policy, strategy and objectives;
- Mergers, business downsizing, business expansion;
- Changes in business processes, operational practices;
- Changes regarding the nature, use and application of information and information systems;
- External changes: political and economic, markets, customer base, environmental, social;
- High turnover of human resources;
- Changes to legislation and regulations that the organization needs to consider to avoid the risk of noncompliance;
- New, or changes to existing, contractual obligations, SLAs or contracts that the organization needs to consider to react to avoid the risk of noncompliance or breach of contract;
- Changes to the supply chain driving the organization or by being driven by the organization;
- Changes in technology;
- Changes in the overall risk profile due to internal or external events and activities.

Again, in all these examples of change there could be information security requirements, risks and impacts that need to be considered.

8.1.4 Tracking and Reviewing Ongoing Change

The overall ISMS monitoring process is about getting enough information and details about all types of change and whether there is a rise in the risk and impact levels. Monitoring the result of change may appear through tracking incidents and looking for trends and patterns: whether there has been an improvement or deterioration in the information security performance.

Several processes can be used to provide information about changes; an obvious one is the incident handling process, which provides information about changes to risks and impacts, and the organization's ability to respond to these. Reassessing the risks and impacts and doing gap analysis also provides similar information.

Scorecards, staff feedback and similar-type human methods can and do provide some valuable information. Technology in the form of firewalls, intrusion detection system (IDS) devices, audit trail applications, various monitoring tool for networks, services, web browsing and web-based applications can also add considerable value in this information-gathering process, all the time building up a better picture of the changes, the trends and patterns allowing future planning and well-informed management decision-making. The organization should not ignore the valuable information that could be available from customer and supplier feedback.

8.1.5 Informed Decision-Making

Informed decision-making in the organization is essential if the ISMS is to be effective, adequate and suitable to meet business needs. The organization needs to have a core knowledge of its strengths, weakness, opportunities and threats regarding its ISMS in the context of its business. This includes having a clear knowledge and understanding of its business and legal requirements in relation to its ISMS, the performance requirements in terms of the effectiveness, adequacy and suitability of the ISMS, and the nature and degree of the risks and impacts it faces, its risk criteria and how much risk it is prepared to tolerate and accept. Having a core knowledge and understanding of its strengths, weaknesses, opportunities and threats will enhance informed policy and decision-making, and this greatly assists its ability to manage its ISMS business support and resources.

As mentioned earlier, information security is more of a management, people, process, policy and procedural issue, and implemented and treated this way, it is a business enabler, to support its business investments and opportunities.

The management of an organization is responsible for the management of the risks its faces. This includes operational responsibility as well as providing assurance that the security processes themselves are suitable, adequate and effective, including the risk management process, the incident handling process and backup process. Risk management, in general, covers the assessment of the risks, the implementation and deployments of controls and processes by which the risks are managed and maintained at an acceptable level. The system of controls (policy, procedures and processes) deployed by the organization should strictly depend on the considered risks and impacts. In order that the right security processes, policies, procedures and other controls are suitable, adequate and effective, informed decision-making is essential and critical to the management review process (Section 8.6).

8.2 Monitoring and Operational Reviews

8.2.1 Monitoring

8.2.1.1 Monitoring and Reviewing the ISMS

Monitoring the ISMS is one of the major areas of requirement and activity in ISO/IEC 27001 (Clause 9.1), which falls under the heading of performance evaluation. There are several aspects of the ISMS that an organization should be considering, including the monitoring and review of:

- ISMS performance in terms of its effectiveness, adequacy and suitability;
- Changes, risks and impacts and their effect on the ISMS performance;
- ISMS processes and information security controls to check their effectiveness, adequacy and suitability, including processes for incident management, the risk assessment/treatment, information processing, access control and human resource management;
- Staff awareness, competence and utilization of the ISMS;
- Efficiency, efficacy and effectiveness of the IT, network services and infrastructure;
- Management of supplier relationships, services, contracts and SLAs;
- Compliance to the organization's policies and procedures, contractual obligations, laws and regulations.

8.2.1.2 Monitoring and Reviewing Performance

The organization needs to monitor and review the performance of its ISMS effectiveness, adequacy and suitability:

- Effective: Is ISMS successful in producing a desired or intended result?

- Adequate: Does the ISMS deliver an acceptable quality and amount of information security?
- Suitable: Is the ISMS right and appropriate for the organization's needs, purpose and nature of business?

Monitoring and reviewing the performance of the ISMS help the organization to maintain and improve the ISMS in terms of its effectiveness, adequacy and suitability. This also means monitoring and reviewing the ISMS in an organizational context (Sections 3.2 and 3.3), that is, the ISMS is effective, adequate and suitable in dealing with both internal and external issues and business context, and the ISMS meeting the needs and expectations for interested parties.

8.2.1.3 What, When, How

The organization should have a monitoring programme that will cover what is to be monitored, when monitoring should be carried out and who is responsible for monitoring. The organization needs to define how the monitoring is carried out: the monitoring methods and techniques to be used and any special practices that need to be considered. It is important that the process of monitoring produces results that are comparable and repeatable to ensure the analysis and evaluation of the results is consistent, defendable and valid.

8.2.1.4 Monitoring and Reviewing Business Demands and Requirements

It should be clear by now that any activity involving the ISMS needs be done in cognizance and familiarity with business demands and being able to meet these demands. This should be the true essence and ethos behind the organization's implementation of its ISMS. Any reviews relating to the ISMS needs to always bear this in mind.

One way of making sure that activities are in synchronization with such demands and the meeting requirements of interested parties is to make information security a topic on the agenda of management meetings at all levels, a part of management decision-making, part of the overall business culture and way of thinking and an integrated part of other relevant business activities and process. This is to ensure that those dealing with the ISMS do not lose sight and focus on the business goals and requirements and senior management are fully aware of the ISMS activities and developments and that continued support and commitment are assured.

8.2.2 Monitoring and Review of Staff Awareness, Competency and Use of the ISMS

8.2.2.1 Competence

Staff competence is generally considered to be that knowledge, skills and behaviour that the organization requires of an individual to properly perform a specific job function or role. For instance, a sales and marketing management role might require the skill in influence and negotiation, whereas a network administrator's role should include technical knowledge and analytical skills. Someone working in the personnel group responsible for all the staff records needs to have a working knowledge of the legislation that applies to the handling of personnel data, information security to protect data, they will need knowledge of handing personnel problems, skills at handling a range of different personnel types, and having good communication skills, to name a few of the necessary competence requirements.

It is often said that a part of human nature and psychology tends to make people self-examine themselves and either underestimate or overestimate their competence: those who are incompetent at tasks sometime overestimate, whilst those very competent people sometimes underestimate themselves. Good management should always motivate rather than demotivate their staff as staff are one of the most important assets the organization. Providing the good management leadership and right motivation, giving the right amount of responsibility and incentives and enabling good communications channels between management and staff are a pathway to success.

Monitoring staff resource levels, their skill sets, their career development and the need for training and retraining is a major task that management needs to undertake (see Chapter 5).

8.2.2.2 Security Culture

Staff are a significant driver, for not only their own productivity, but the collective productivity and efficiency of the business; therefore, it important to have both good human resources practices in place as well as instilling a good security culture. Developing a healthy security culture within the organization should be one of the main aims of the organization. This topic of security culture is so important as a management control for information security it is being addressed by various international institutions and organizations, including the Organization for Economic Co-operation and Development (OECD) in their article, "Guidelines for the Security of Information Systems and Networks: Towards a Culture of Security," July 2002.

Good human resource management should be there to deal with internal staff problems or conflicts, for example, dealing with disgruntled or dissatisfied staff, those operating under poor working conditions or under undue pressure or those who are not being used to their levels of skill and competency. Dealing with these issues as soon as possible can avoid some of the insider risks mentioned earlier, whilst supporting staff productivity and efficiency.

Some organizations have had a security culture operating in their organization for many years. These organizations have realized the importance and benefits that such a culture can offer to the overall awareness and subsequent information security of their business as a whole. It is often said that IT is not the problem of information security; it is the people using the IT, which matches up to many of the security surveys which report that people are the cause of many security breaches and incidents.

8.2.2.3 Monitoring Staff

Monitoring and reviewing staff in relation to their responsibilities and roles regarding information security are extremely important. However, this should be done as part of the normal line management and human resource processes that already exist. It should not be done in a Big Brother way, as this will make staff feel ill at ease, demotivated and overly restricted, which can lead to mistakes or accidents in their work. It might have an impact on their efficiency and efficacy, and it could, in some cases, lead to a disgruntled employee launching an insider attack. So a balance needs to be reached between a laissez-faire and a draconian style of human resource management.

Many organizations have adopted Acceptable Use Policy (see Section 6.4.2) to control the use of organizational resources and facilities, for example, IT, internet services and email.

All forms of monitoring and review of staff activities needs to be done in accordance with local laws and regulations.

8.2.2.4 Awareness, Training and Career Development

Developing and improving staff knowledge and skills requires an appropriate level of training and awareness. Formal classroom training, on-the-job training and online training and learning are some of the ways the organization can utilize to help their staff in attaining the right levels of knowledge and skills for the job. As people are of such an importance to an organization to meet their objectives and performance levels, it is equally important to look after staff training, awareness and career development. Not only does this ensure that the company has the right competent people to do the job, but it also motivates staff as their career development is being looked after and avoids unnecessary loss on skilled staff that does not receive such attention. Section 8.3 covers example measurements concerning staff awareness and training.

Monitoring staff regarding their ongoing competence to perform specific job functions enables the organization to further develop the knowledge, skills and abilities of staff to continue to perform such a job efficiently, effectively whilst protecting the organization's information.

The organization might want to transfer the individual to another job function requiring a different set of competencies, and monitoring the individual's ability, knowledge and career development enables the organization to provide that member of staff with retraining. The human resource group plays a key role in managing the process of matching competence requirements of the job to the skills and knowledge of existing staff or the recruitment of new staff.

8.2.2.5 Some Specific Information Security Roles

The following is an example list of information security roles related to ISMS:

- Chief Information Security Officer (CISO), information security officer(s), manager(s);
- Information security-related management roles and tasks or job functions regarding information systems, human resources, operations, service management and business continuity;
- Information security risk manager(s);
- Information security incident management team;
- Internal ISMS auditor(s);
- Compliance manager(s), PII officer;
- IT and network-related security roles.

Everyone in the organization will have some part to play in information security from top or senior management through the various levels of management, as well as the nonmanagerial staff.

8.2.3 Monitoring and Review of Information Security Processes

8.2.3.1 General

In ISO/IEC 27001, many processes are defined, some of them at the management, auditing and compliance levels, as well as at the human resource level and at the technical level. These processes will need to be monitored and reviewed to check their performance, and if the performance seems to be below standard, then the process needs to be improved. For example, if a backup fails to function properly, does not complete properly or does not deliver according to what is expected, data recovery is difficult or unsuccessful or the process generates errors on a regular basis or many other possible problems, then the process needs to be reviewed and improvements to the process are necessary. Other examples of monitoring include the staff recruitment process, the user access management process (granting and removal of accounts, allocation of access rights, privileges and passwords and authentication attributes), service management process or various activities relating to information processing.

8.2.3.2 Risk Management Process

The risk management process itself should be reviewed from time to time. The effectiveness of the ISMS is dependent on quality of the risk management: monitoring and reviewing the risk assessment results, the assessment criteria and methods used, the method for determining which controls are to be implemented, the risk treatment decision-making and the calculation of the residual risk, all these should be part of the monitoring and review exercise.

8.2.3.3 Risk Monitoring

The risk environment of an organization will be in constant change and, as a consequence, the risks and impact to an organization need to be monitored and reviewed. The following factors are some of those the organization needs to monitor and review since they may increase or decrease the risks associated with the ISMS:

- New threats and vulnerabilities;
- Increase or decrease in information security incidents;
- New assets added to the scope of the ISMS or current assets changed;
- Changes to business processes and/or operations;
- Expansion, restructuring, downsizing of the organization;
- Changes in workforce;
- New services deployed;
- Changes to outsourcing arrangements;

- Introduction of new technology;
- External changes in the business environment;
- New legislation or regulations or new contractual obligations and arrangement.

The results of the ISMS risk assessment we considered in Chapter 4 need to be regularly reviewed and reassessed because of the potential changes due to any of the factors above.

8.2.4 Monitoring and Review of Information Security Controls

In ISO/IEC 27001, many requirements are defined and designed to serve different purposes: management, human resource, policies and procedures, auditing, compliance and technical. Again, regular monitoring and review of controls enable the organization to keep track of its performance and to take appropriate action to implement improvements to the controls as and when necessary. For example, an acceptable use policy may be badly written, with instructions, of what is and is not allowable behaviour with regard to email usage, being vague and confusing. Several incidents have occurred whereby users have accidently misused the email system, with the root cause seeming to be to be this badly written policy. Another example might be monitoring the level of resources needed to deal with the day-to-day management of some of the ISMS processes. If we consider the incident management process as a case in question, and the level of team resource needed to do the work, it might be that the number of incidents is steadily increasing, or the organization is experiencing very high bursts of activity. This might demand extra skilled support for the existing incident team on either a full-time basis or a part-time basis, whatever the situation monitoring demand for ISMS human resource is will be important and, in some critical cases, could mean the difference between survival and complete disruption and failure. Monitoring the needs and requirements across all areas of ISMS activity is important.

There are many management requirements in ISO/IEC 27001, including the management control of the policies and procedures, all of which can be the subject of some form of monitoring. Documentation, such as policies and procedures, will be created, approved and used in day-to-day operations, and, at some point in time, they need to be updated and reviewed; this is normal practice and part of ISO/IEC 27001 compliance. Monitoring is highly necessary to check that all ISMS-related documentation is suitable and adequate, is available to all those who need to have access, is suitably and effectively protected, has adequate suitable version control and is adequately and effectively stored, archived and disposed of as is necessary.

Some determined controls, such as those that are underpinned by the use of IT, may be monitored using automated or semiautomated means, for example, software used for monitoring unauthorized access or detecting intrusions or for generating audit trails.

8.2.5 Monitoring and Review of Technology and Network Services and Infrastructure

ISO/IEC 27001 is not an IT security standard, it is an information security management standard: it is driven by management and the organization and not by the IT department, it deals directly with the security of information and the risks and protection of this information and not with IT security, network security and the security of other technologies. Nevertheless, technology today is a significant management tool and as such, like other management support and business tools, methods and processes, technology has implications on the effectiveness, adequacy and suitability of the organization's ISMS and the ISMS outcomes the protection of information.

Monitoring and reviewing the use and application of technology within the ISMS environment, the following are some of the things that need to be considered:

- User and management policies and procedures that include the secure and acceptable use and application of technology; user access management and secure information processing involving the use of technology; and business communications and transactions using network services;
- Information security management aspects related to the use of business, operational and application software;
- Information security management related to the use of technology for business processes, decision-making and the evaluation and assessment of market demands;
- Information security incident management related to the use of technology for the detection, evaluation, response and recovery of incidents and for incident forensics;
- Information security management aspects related to the use of technology and infrastructure for business continuity, availability and resilience.

8.2.6 Monitoring and Reviewing Third-Party Contracts and Services

8.2.6.1 Supplier Relationships and Service Management

The majority of organizations have third-party suppliers providing a range of different services. Management of supplier relations and service delivery especially should consider involving information security.

The organization should have a policy in place that covers the information security risks with regard to the use of suppliers and the protection of the organization's information.

Also, information security should be addressed in contracts and SLAs with the supplier to ensure that the organization's requirements are being addressed.

Both the policy and the contract or SLA should address the risks related to the supply of services and products.

The information security policy, contracts and SLAs should be revised on a regular basis and more immediately if there have been any changes or requests for changes (reduce or expand its current services, revise the current level of service or add new services). Another reason for revising the contract and SLA is if there have been service-related incidents that would require a reassessment of the risks.

Monitoring and reviewing of third-party services might include checking and evaluating records of service performance, service levels, and any problems in not achieving the requirements of the SLA, risk levels, service recovery and any incidents that may have occurred.

Clearly, the organization's risk assessment and risk treatment solutions regarding service management needs to be reviewed on a regular basis. In addition, monitoring the levels of information security risk and impact needs care and attention to ensure that the ISMS still delivers the desired results effectively.

Another aspect is the auditing of the supplier's system to check the compliance against what has been agreed between the contracted parties. The right to audit the supplier's system is something that needs to be agreed between all parties involved: the client and the supplier(s). Such an agreement needs to be included in the service contract, and this might stipulate various conditions and restrictions. Some organizations have included this right to audit as part of the tendering process and some organizations have made third-party certification audits (Chapter 10) a condition of the tender.

8.2.6.2 Monitoring and Reviewing Service Provider Changes

Any changes on the supplier side need to be communicated to the organization so it can review whether these will impact it. There will always be a need for IT changes at the service provider end. Maybe their success has secured them more business with existing clients or new clients to meet the increase in demand for IT facilities and services. It might be the case that the provider wants to make changes to stay ahead of its competitors and to be number one in its field of outsourcing, replacing outdated systems or by introducing changes to improve, streamline and/or make more cost-effective its operations. Whatever the reason, IT changes might involve all or some hardware, software, operational environments, management tools, networking arrangements, connections and services to change. Conveying these changes to its clients through regular channels of communications and reporting help both sides to reassess the risks and plan for any changes that might directly affect the client. This type of reporting should be in the existing contract and a contract review process needs to be initiated to discuss, negotiate and agree any amendments or modifications to the contract.

8.2.6.3 Termination of Contract

There are several reasons for terminating a contract: the supplier is unable or not willing to expand its services, costs and value for money or lack of information security. It might be that the service provider is taken over or merged with other suppliers, which might present a greater security risk or simply that the service goes out of business, or the supplier has decided to subcontract the work to a supplier located in a high-risk area. It might be that the organization switches supplier for information security reasons.

Whatever the scenario, there are several information security aspects to be considered. If a termination does go ahead, how is the organization going to protect its information assets between finishing one contract and starting another? Does it have the skills in-house to protect and manage itself during the transition? The organization needs to take of the fact that, in any transition between suppliers, all parties need to be involved to ensure a relatively problem-free changeover, especially to ensure that no glitch in the continuation of its operations.

The business continuity and availability issues need to be addressed with regard to a change in suppliers or providers, to ensure that during the transition period there is adequate, effective and suitable protection of an organization's information assets.

8.2.7 Monitoring and Review of Legal and Contractual Compliance

ISO/IEC 27001 specifies that the legal and regulatory requirements pertained to the ISMS need to be considered. This also includes consideration of contractual requirements and obligations. It is vital that an organization monitors, reviews and keeps track of its compliance with those laws and regulations that apply to it, reviewing and reassessing the risks and making changes to the ISMS where relevant and appropriate.

With multinational organizations, there is potentially more legal and regulatory requirements of which to keep track of and with which to comply with. In addition, when trading across the internet between jurisdictions, there may be several laws and regulations that might apply.

In all cases, the organization needs to make sure that its ISMS implementation, like all other aspects of operations, should be complaint with all relevant laws and regulations. The impact, for example, in terms of legal penalties, fines, legal punishments for individual company officers and damage to image, as failing to comply can be very severe. It is therefore essential that the organization knows with what it should comply and what controls it needs to implement in this respect as part of its conformance with the ISMS requirements.

8.3 ISMS Measurements Programme

8.3.1 ISMS Metrics and Measurements

8.3.1.1 Metrics and Measurements

A metric is a system or standard of measurement, and a measurement is the result of measuring something against a metric. For example, we have the litre as a standard for measuring the capacity of liquid, seconds as a measurement of time or degree Celsius as a measurement of temperature.

In terms of information security, there are many types of metrics or measures that can be defined, for example, the number (maybe as a percentage) of system vulnerabilities that are resolved within a defined time after set discovery time or the number and percentage of incidents that are resolved within a defined time after set discovery time. Another example might be the effectiveness of the security help desk: number and percentage of calls received, answered or abandoned, average query processing time and average waiting time, number and percentage of resolved and unresolved problems and user or customer satisfaction. Whatever metrics are selected, they need to be appropriate to the organization's requirements to deliver and maintain an effective, adequate and suitable ISMS.

8.3.1.2 ISMS Measurements

In terms of the ISMS, the organization needs to define what metrics and measurements that it needs to use, and how, when, where and by whom or what measurements are to be taken. The selection of metrics (measures) is entirely the decision of the organization, as is the how, when, where and what regarding the measurements. It must always be focused on the aim of taking measurements (why do we need to take measurements), for example, to check the performance of the ISMS with regard to its effectiveness, adequacy and suitability, and these measurements provide valuable input with regard to the monitoring and review of the ISMS and consequently helps to determine what improvements need to be made to the ISMS.

Taking measurements and analysing them are mandatory requirements of ISO/IEC 27001 if an organization wishes to claim conformance with this standard. In addition, taking measurements has many business benefits for the organization, for example, it greatly assists in managing the risks and the allocation of resource and budget towards controlling the risks. It also helps to demonstrate conformance, that is, the organization is meeting the needs and expectations of interested parties. It also helps to increase accountability and to check ISMS implementation in operation. If an organization invests in developing and implementing an ISMS, then it wants to see some return on its investment. Is my ISMS performance as intended and does delivery provide the right outcomes? Monitoring, measuring and evaluating its performance help in answering this question.

8.3.1.3 Examples of Areas of Measurements

Measurements can help to indicate the extent to which the ISMS is delivering the desired and expected results (effectiveness), the extent to which the ISMS it is achieving an acceptable level of quality with regard to information security protection (adequacy) and the extent to which the ISMS is right and appropriate for the organization (suitability).

The following are some example areas to illustrate possible ISMS measurements.

Management processes are:

- Effectiveness, adequacy and suitability of the process for the ongoing determination of the requirements, needs, expectations for and issues related to the ISMS;
- Effectiveness and adequacy of ongoing leadership and support for the ISMS;
- Effectiveness, adequacy and suitability of internal and external communications on all issues, news, changes and improvements related to the ISMS;
- Effectiveness, adequacy and suitability of the ISMS documentation system, including its protection, completeness, currency, availability, distribution, changes and updates.

Human resources are:

- Operational performance of staff and their knowledge and understanding of security procedures;
- Effectiveness, adequacy and suitability of user training and awareness;
- Effectiveness, adequacy and suitability of human resource security: recruitment, during employment and termination of employment.

Security processes are:

- The effectiveness and suitability of the risk assessment and treatment processes;
- The effectiveness and suitability of the information security incident management process;
- The effectiveness and suitability of the backup process; vulnerability management process; configuration and change management process; maintenance process;
- User access control management process:
  - User accounts (register and removal);
  - Managing (granting, changing and removing) rights, passwords and privileges and authentication attributes.
- Effectiveness, adequacy and suitability physical and environmental controls.

Management and user controls are:

- Levels of usage and demand for information processing, IT and network services resources to help with capacity planning;
- Effectiveness, adequacy and suitability of policies and procedures:
  - Access control;
  - Backups;
  - Acceptable use;
  - Media handling.
- Access methods (to information, applications, systems, networks).

Reviews and audits are:

- Effectiveness, adequacy and suitability of ISMS audits, including specifics such as number of observations and noncompliances found, whether major or minor noncompliances, closure rate for noncompliances;
- Gap analysis and reviews of process and control implementation: checking the gaps and weaknesses in the implementation;
- Effectiveness of management reviews.

Economics, investments and risk or impact costs are:

- Information security budgets and resources;
- The cost-benefit analysis regarding investing in information security versus impacts;
- Incident management, disaster recovery and business continuity costs; costs of recovery, cost of replacement, legal costs, contractual penalties, costs against impact analysis and other costs with information security dependencies.

8.3.2 Measurement Programme

8.3.2.1 Programme Design Objectives

An ISMS measurements programme should be designed to provide the organization with:

- A valued assessment as to how well they are doing and how well the ISMS processes and controls are being utilized and how well they are performing; set benchmarks and performance indicators to be set;
- Better informed decision-making about the ISMS effectiveness, adequacy and suitability; risk treatment solutions, making improvements;
- Demonstrate to CEO, directors, senior management, other interested parties and stakeholders and the board the value of the ISMS in the sense that risks and impacts are being managed effectively to achieve the desired ISMS results, there is an adequate level of protection provided by the ISMS, and the ISMS is suitable to meet the organization's needs, and that it remains so.

Taking measurements should be seen as an activity of great importance towards achieving, maintaining and improving ISMS effectiveness, adequacy and suitability. The measurements activity needs to be link with the monitoring activity, as measurements should be taken on things that are being monitored for effectiveness, adequacy or suitability. Therefore, all that was said in Section 8.2 applies here in Section 8.3. Equally monitoring and measurements are tied into the management reviews, as we will discuss in Section 8.6.

One way to interpret the requirements in ISO/IEC 27001 Clause 9 on Performance Evaluation is to say that the organization needs to:

- Have a set of metrics (predefined standards of measurements: types of measure) against which to take measurements.
- Define the information security targets and thresholds, key performance indicators and how these targets are calculated, reviewed and updated, bearing in mind that measurements should be objective, reliable, accurate, repeatable, verifiable and relevant and quantifiable against the business targets.
- Have a set of processes, procedures and methods to take and evaluate these measurements.
- Define the frequency of taking measurements: when measurements should be taken; and which staff members are allocated the task of doing measurements and who needs to ensure they are taken (bearing in mind that some measurements could be manual, semiautomated or automated).
- Define when do the measurements need to be checked, assessed, analysed, evaluated and reviewed; and which staff members are allocated this job function.

8.3.2.2 Measurement Methods

The standard ISO/IEC 27001 does not define any measurement methods, as the number of ways of taking measurements is numerous. Also, a measurement system is very dependent on the organization's internal standards, processes, mode of working and business culture. It is likely that the organization will have performance evaluation exercises in other areas of their business, for example, project management, quality management or problem management. The methods, such as incremental milestones, cost ratio and expert and customer feedback, used in these other areas for performance evaluation might well be adopted for ISMS use. Again, this depends entirely on the organization and its senior management needs to decide which methods, metrics and measurement system it should adopt to satisfy the requirements of ISO/IEC 27001.

There are some data management methods that do apply to the process of measurements and can be used for the gathering and collection of measurement data: and some methods can also help in checking the quality, integrity and validity of the measurement data. It should be emphasized that it is up to the organization to decide and manage its measurements systems and which methods are adequate and suitable for its own use. It should also be emphasized that the monitoring and measurement system and methods used are themselves a subject for performance evaluation. This means that whatever metrics and methods and the who, when and how of taking measurements also comes under the scrutiny of being assessed and reviewed, as improvements may well be necessary if the monitoring and measurement system that this is adopted is itself not performing well, that is, it is proving to be ineffective, not adequate enough or unsuitable. The monitoring and measurement system needs to be reviewed at planned intervals to check its effectiveness, adequacy and suitability.

8.3.2.3 Protection of Data

It is important to note that some of the data being collected as part of the monitoring and measurement exercise may contain details of real or potential weaknesses or vulnerabilities in the organization's operational systems and processes. As such, this type of sensitive data, in the form of records and other documentation, needs to be adequately protected to avoid leakage, unauthorized access and unauthorized modification.

8.3.2.4 Data Gathering and Collection

In this clause, we will present here is some of the basic methods of gathering information for measuring performance. In general, an organization only gets "what they measure in terms of targets and indicators," so if they do not measure the right things, they do not set the most suitable performance targets and indicators or ask the most appropriate questions and they may not arrive at the right answers to tell them how well their ISMS is performing. This is easier said than done; hence, the reason to use continual improvement, which helps to monitor, review and improve processes, including those used for measurements. Therefore, care and attention in the design of these methods and measurements will avoid unnecessary work and provide more appropriate decision-making results.

Additionally, the quality, integrity and validity of data depend on the sample taken, for example, a poor sample, a small sample or a sample that is not truly representative of a particular aspect of the ISMS. If the results of the measurement programme are to be useful, adequate and suitable to be of benefit to the organization, then the data needs to be collected from a truly representative, statistically significant, consistent sample.

Another important thing to bear in mind is that the exercise of data gathering or collection and measurements thereof should not interfere with business operations.

As already mentioned, data gathering could be manual, semiautomated or automated.

8.3.2.5 Measurement Templates and Example

Some organizations use a standardized set of templates for recording measurements. Each type of measurement would have its own predefined template covering: a definition of the metric being used, the ISMS objective of the metric, details of the measurement to be taken against this metric, the type of measure (e.g., measure of effectiveness, adequacy and/or suitability), the performance target associated with this measurement, the range and type of data to be collected for this measurement, the frequency of measure, who is the owner and other interested parties for this measurement.

For example, a measure is defined by the organization with regard to the effectiveness, adequacy and suitability of its ISMS training and awareness programme: goals are to have a staff with knowledge, working experience and operational competence in the use of ISMS processes and controls and that the training they have received is effective, adequate and suitable for their job function and to carry out their job function and their related roles and responsibilities; the measure is a percentage of effectiveness, percentage of adequacy, percentage of suitability as determined by a test, examination and operational demonstration; the percentage targets are set by the organization; the measurements are carried after each planned training session (e.g., quarterly); the interested parties are human resources, head of training, training managers and line managers.

8.3.2.6 Incident Management and Other Reports

Incident handling reports provide a wealth of data as regards the state of the ISMS to protect the organization's information. These reports will have details covering what the incident problem is, its nature and characteristics, how and why the incident happened, which systems and process were affected and what was needed to recover from the incident. Incident reports over a period of time provide data on the trends and patterns of incidents and their type and nature. This trend information provides valuable measurement data, for example, unauthorized access attempts, how many attempts have there been and how many were successful, access to which part of the organization and its systems and processes was attempted, types of damage incurred, which systems or processes were compromised and other types of information. All this data and information, of this incident and all other incidents, provide indications as to whether the ISMS provides adequate protection, whether this protection is effective and whether this protection is suitable in the context of the organization's business and the needs of interested parties. It is therefore essential that the incident management documentation and records are kept since this system will contain substantial volumes of data which the measurements programme could usefully use.

8.3.2.7 Scorecards and Questionnaires

Scorecards (especially balanced scorecards) provide a means to measure the performance of key business processes against business strategy and objectives. For example, the scorecard could be used to measure the performance of the incident handling process against the ISMS objectives referred to above. This might consider score questions related to:

- Incidents reported and resolved over a period time relating to unusual website behaviour, adware, spyware, phishing and other similar issues across different parts of the organization, how long did it take;
- Time spent recovering from system downtimes, failures and/or unavailability of network services;
- Time spent in correcting user errors in data entry and processing activities.

The results of using scorecards provide management with a snapshot of how effective their risk management controls are. It is a tool that can be used alongside other methods that have been mentioned above to complement the contribution of information gathering during the monitoring and reviewing process.

Scorecards should be used on a regular or periodic basis to get the most benefit and to support the continuing pursuit of effective information security through the process of continual improvement.

Equally any security questionnaires asking questions about ISMS use and implementation also provides measurement data. If the questionnaire is well designed with targeted questions of particular interest to the measurements programme, then the answers provide another source of useful data.

8.3.2.8 Tests and On-the-Job Exercises

Testing and exams are one method of collecting data to measure performance related to the human resource aspects of the ISMS. These are designed to test the knowledge and understanding of staff with regard to the ISMS they are using. This might be testing the knowledge of a set of policies and procedures with example exercises and problems to solve on typical work scenarios. On-the-job training and at-the-desk exercises to test the use and application of ISMS processes and controls.

Benchmarking is a useful tool as it can provide a means of comparing different parts of the business based on a set of metrics aligned to the objectives, needs and expectations of the interested parties involved in the ISMS. As regards the effectiveness of their information security and their level of compliance and progress. It can also be used to evaluate and check how a company is doing with regard to its information security with respect to other companies working in the same line of business or in the same industry sector.

Benchmarks provide an innovative way of viewing information security as a business enabler across different areas of the business identifying areas for improvement, integration, streamlining and greater efficiency and productivity.

8.3.2.9 Gap Analysis and Benchmarking Methods

Gap analysis is another assessment tool, which enables an organization to compare its actual information security profile with what it needs to be compliant and/or the security objectives and targets it wants to achieve and to compare the actual performance with its potential performance. The organization can use this to check against predefined metrics and targets aligned with the organization's objectives, needs and expectations and that of all the interested parties involved in the ISMS. This provides the organization with measurements of compliance against set targets, for example, the checking the extent of implementation and operational usage of the standard principles of privacy (as set out in ISO/IEC 29100). This gap analysis will provide insight into areas where there is room for improvement.

The gap analysis process provides a means of determining and documenting the variances between the target ISMS business objectives, needs, expectations and requirements and what is actually implemented in the ISMS (Table 8.1). There are many uses of gap analysis:

- Pre-risk assessment to check what controls and processes are currently in place;

Table 8.1
Gap Analysis

Question	Compliance Targets				Comment	Action	Closure
	C	S	P	Z			

- Post-risk assessment, treatment and selection of controls, activities and actions;
- Check status of implementations, actions and status related to management reviews and improvements;
- During ongoing risk management activities such a risk reassessment;
- Pre-certification audits as well as internal audits and reviews;
- Check operational compliance with ISMS policies and procedures;
- ISMS asset management;
- Protection of information life-cycle management;
- Incident management applications;
- Business impact analysis;
- ISMS resource allocation, provisioning and management.

The granularity of the analysis can vary in several ways, for example:

- Targets of compliance complete (C), semi-complete (S), partial (P) or zero (Z);
- The number of questions asked to establish the level of compliance can also vary.

The greater the granularity of analysis, the more precise the results are likely to be. The following example illustrates this idea.

For example, let us assume that the organization carries out measurements on its service and supplier relationships. The purpose of this is to check that the controls in place to protect the organization's assets, including its information assets, are effective, adequate and suitable. A single question to assess whether the organization has an information security policy in place for each of its suppliers might be: Has the organization put an information security policy in place to cover the risks associated with access to its information assets by service providers?

However, this question is very simple and gives no indication of whether policy specifics are complied with. So an expanded set to several questions might be used to give a greater degree of measurement of compliance of the specifics of the policy:

- Does the organization's information security policy for service providers cover controls related to the protection of the confidentiality integrity of organization's information?
- Does the organization's information security policy for service providers cover controls related to the protection of all personally identifiable information that the organization has?
- Does the organization's information security policy for service providers cover controls related to incident management?
- Does the organization's information security policy for service providers cover controls to ensure availability of resources used to process information and of information itself?
- Does the organization's information security policy for service providers cover monitoring the quality of service provided?

8.3.2.10 Automated Data Collection, Processing and Learning

Manual data collection, although inevitable for several measurement exercises, is not necessarily the most efficient. In order to add greater efficiency, the use of automated tools for data collection and processing is far better.

Now there are several ways automation can be used. Clearly having most of the methods mentioned above automated is possible. For example, scorecards, questionnaires and sometimes gap analysis can be done online and so all the results are in electronic form, which makes processing, analysis and future use and application easier.

Other forms of automated data collection relate to reporting of incidents, problems, errors, vulnerabilities and customer feedback; although these may not directly relate to any measurement exercise per se, they are data that is useful in the broader scheme of things and do constitute a measure of the effectiveness, adequacy and suitability of the ISMS.

There are indeed software tools that are specifically designed to collect data, analysis data and present results. IDS software provides indications of traffic and user behaviour against predefined patterns of behaviour, firewalls and other network technologies also provide data related to network usage, behaviour and problems. IT has audit trails built in that provide data on user access to files, applications and services, systems and various other things.

So automated data collection, processing and analysis lend themselves to being an integral part of a measurements programme. However, not all the work that is needed for an ISMS measurement programme can be automated, since much of the ISMS is about people, management and the use and application of information. Putting automation and IT in context, we can say it is an extremely important business tool designed to make processing of information more efficient, but we cannot totally automate the human aspect, from a usage, behavioural and application perspective. The use of artificial intelligence and machine learning can be used for some aspects of data processing and measurement evaluation.

8.3.2.11 Measurements in Practice

There are several key points that should be kept in mind by the organization in designing an effective measurement programme such as:

- Value and purpose:
    - Provides a measure of performance of the effectiveness, adequacy and suitability of the ISMS against the organization's objectives and the needs and expectations of all interested parties.
    - Provides an understanding of possible risks (new risks or changes to existing identified risks) and provides an indication that more controls are needed or existing processes and/or controls need improvements.
    - Helps in monitoring the ISMS and provides essential information for the management reviews.
    - Identify problem areas where improvements could be made.
    - Identify areas where the performance is below par.
- Type, quantity, quality and timeliness:
    - Make sure the measurements are relevant, and a sufficient number of measurements are taken, regularly and at the right time.
    - Make sure you are not measuring too many things that are not used; the measurement programme needs to be focused and consider what is relevant, the critical parts and potential risks of the ISMS.
    - Make sure imprecise measurements or irrelevant measurements

are not being taken.

- Use and interpretation of measurements:
    - Make sure measurements are being undertaken with a clear goal in mind.
    - Make sure all those that are asked to provide input understand the purpose and relevance.
    - Right use (e.g., to gain a better understanding the state of the ISMS).

It is important that whatever metrics are used and whatever measurement system is in place to measure against these metrics that it will deliver meaningful, useful and timely results. There is always room for improvements as this is the nature of change and this is why monitoring and measurements play an important role in the lifetime of an ISMS. No one can get a measurement programme fully correct from the word go and it is unlikely that such a goal in reality can never be attained because of change. Therefore, the measurement programme itself needs to be reviewed, especially the relevance of the metrics being used, the method of taking measurements and frequency of measurements.

8.4 Ongoing Risk Management

8.4.1 Risk Responsiveness and Commitment

At the heart of this, like all other aspects of the ISMS, is the topic of risk; we cannot get away or remove ourselves from this important and critical topic. Measurements might inform us that there is a problem with the ISMS, that is, there are risks that are causing this problem: new risks, variants of old risks or an old risk that has become more virulent or hostile or more likely carrying a greater impact, in which case the organization needs to reassess its risks. At the same time, a risk assessment, in and of itself, is a measurement of how good the ISMS is in terms of effectiveness, adequacy and suitability.

Taking suitable action and being responsive to manage the information security risks an organization faces are core to achieving an effective, adequate and suitable ISMS. The processes of risk assessment, risk treatment and selection of controls constitute the first phase of the risk control procedure (as discussed Chapter 4). The second phase compasses implementing the controls and supporting actions to put in place a system of controls and getting them deployed in the ISMS (as discussed Chapters 5, 6 and 7), and the third phase involves keeping ahead, being present with the latest changes, responsive to changes where appropriate, and maintaining an effective information security regime (the topic of this chapter).

At all stages of the risk management process, there needs to be a firm commitment from senior management, operational managers and staff. All have a part to play to counter the threats and incidents that result in unwanted business risks and impacts. Table 8.2 shows a simplified view of typical levels of commitment and responsibilities.

8.4.2 Regular (Planned) Risk Assessments

From time to time, the organization's risk profile needs to be reviewed and updated. This is essential to keep up the changes, to make improvements and to maintain an effective, adequate and suitable ISMS implementation. It is a key part of the continual improvement process (Chapter 9).

This means that risk assessments should be carried out on a regular basis. In addition, assessing the risk will also be needed in the case of a major incident, a change in the business or management review (unplanned risk assessment); see Figure 8.1. This will include reviewing the current status of

Table 8.2
Levels of Commitment and Responsibilities

Human Resource	Level of Commitment and Examples of Responsibilities
CEO, directors, senior management	Commit funding, recruitment of staff and other resources for the management of information security risks, ISMS establishment, implementation, monitoring, reviews and measurements and the continual improvement. Set corporate policy, strategy, and business objectives for risk management activities. Establish risk criteria. Ultimately accountable for the management and control of the risks and impacts. Establish and visibly support an information security culture, including training and awareness. Define and allocate roles and responsibilities.
Operational management and other areas of middle management activities (such as human resources, administration)	Implement policy and business objectives in the form of operational risk management policies and procedures. Measure operational performance, effectiveness and efficiency. Manage operational staff and their deployment of risk controls. Recruit staff. Set operational risk control responsibilities providing appropriate staff awareness; training and on-the-job training where necessary. Monitor, review and report back to senior management risk-related issues, incidents and risk management performance.
Staff and users	Be aware and responsible for information security risk management in their area of work, and understand information security as an integral part of the work. Report security incidents and feedback on effectiveness of security procedures. Be responsive and work as part a collective team throughout the organization to manage the risks.

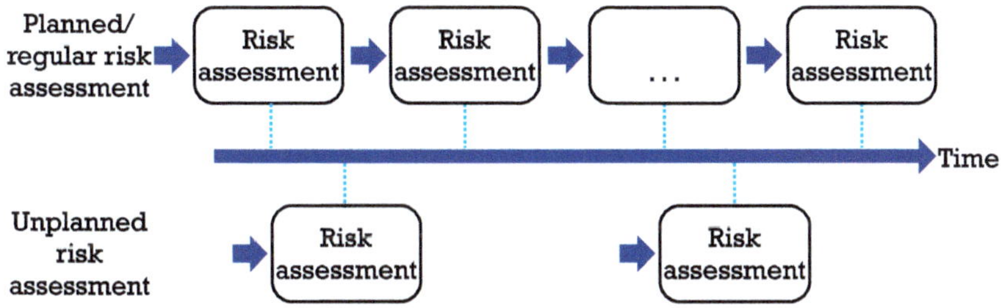

Figure 8.1 Risk assessment frequency.

previously identified risks and impacts as well as considering new risks and impacts. The risk register will then need to be updated. As an example, this would typically include:

- Previously identified risks and impacts;
- Reviewing, reassessing and recalculating previously identified risks and impacts, checking levels and status appropriateness;
- Adding, modifying and/or deleting entries or details of all relevant changes in the risk register;
- Newly identified risks and impacts;
- Assessing newly identified risks, taking account information contained in incident reports, audit reports, security reviews, customer, user, management feedback and reports, any changed conditions and any new risk relevant information;
- Calculating the impacts related to the new risks;
- Reassess the risk priorities.

8.4.3 Risk Measurements and Metrics

Different types of metrics and measurements are available that relate to the risk management-based process.

Quantitative metrics and measurements are based on the use of statistical or historical data and monetary values. This includes assets where real quantities can be used to associate damage directly with some numerical value:

- Cost of recovering from a system failure, virus attack, denial of service and other incidents;

- Cost of replacing a damaged system, piece of equipment;
- Reliability figures of equipment or component failures (e.g., MTBF (mean time before failure));
- Records relating to system downtimes, access attempts and network data (e.g., transmission delays, access delays, call and session drops, error rates);
- Statistics relating to earthquakes, tsunamis, volcano eruptions, flooding, typhoons and other natural threats and pandemics.

Some assets are difficult to assign a quantitative risk measure to (e.g., information itself) or company image and reputation, the skills of human resources or the competence of the attacker and their capacity and resources of carrying out the attack. Therefore, quantitative methods for ISMS risks and measurement are not always possible to apply due to the lack the data available, in which case qualitative methods are normally applied.

Qualitative metrics and measurements do not involve the use of statistical or historical data but use a grading system such as low, medium, high and very high to express the value of the risk, impact or likelihood (see Clause 4.2.2.4, Figure 4.1 and Table 4.1). Although this may seem to be an imprecise method, it is generally a good method to deal with the broad range of object being dealt with in the ISMS, people, management, process and information. Hence, qualitative methods are generally more suited to an ISMS because there is often more uncertainty in assessing the information security environment. In practice, the complexity of information security issues requires a combination of qualitative and quantitative assessment, especially so as the assets will cover tangible and nontangible, management and nonmanagement asset types.

Semiquantitative methods address the need to express qualitative measurements in the form of numbers (e.g., low risk = 0, medium risk = 1, high risk = 2 and very high risk = 3). The numbers themselves can be further quantified in more meaningful terms, such as monetary values, degradation in levels of service and availability, reduction in performance levels and decrease in efficiency and productivity output.

8.4.4 Risk Management Reviews

Reports and records of previous risk management processes are important when reviewing the performance of the ISMS. Reviews provide the organization with an indication whether or not the process is effective and enables improvements to be made to the process as needed.

The purpose of risk management reviews is to provide the organization with some confidence that the risk management process is working correctly, thereby giving assurance that the risks are being managed effectively. This in turn provides a certain sureness which helps the organization achieve its objectives and provide informed decision-making. The review reports should at least document:

- Status of key risks surfaced through the process highlighting any material changes that might modify their likelihood of occurring and the impact, and any breach of the organization's risk appetite in terms of thresholds, tolerance or limits;
- Status of actions mitigating for key risks, indicating where actions are closed, still in progress or behind schedule;
- Significant emerging risks.

8.4.5 Information Security Incident Management

Information security incident management reports are a valuable source of information with regard to the current level of ISMS protection. An incident report might indicate, for example:

- New weaknesses or vulnerabilities associated with the ISMS not seen, reported or assessed before;
- New threats not experienced before or an increase in the threat level;
- Ineffective implementation of a security control or measure or defects in the controls;
- The incident impact analysis show that the potential risk level has increased.

Post-incident impact analysis can provide useful feedback and input to a risk assessment exercise. Consideration of incident records over a period of time is also valuable input to the risk assessment exercise.

8.5 ISMS Internal Audits

There is an important requirement in ISO/IEC 27001 (Clause 9.2) that the organizations need to conduct internal audits. The purpose of an internal audit is to check whether the ISMS satisfies the requirements of ISO/IEC 27001 (Clauses 4–10). This will consider the effectiveness, adequacy and suitability of the ISMS. This audit needs to check for any weak links, gaps and nonconformities in the implemented ISMS against the requirements of ISO/IEC 27001. The audit should also identify any potential opportunities for improvements to correct and fix any nonconformities.

These audits should be carried out at planned intervals as part of an ISMS audit programme. Each audit should have an audit plan, which details the objectives and scope of the audit, the audit criteria. The objectives of the audit are what needs to be achieved by the audit, including conformity of the ISMS with the audit criteria and to determine whether the ISMS meets the relevant legislative, regulatory or contractual requirements. The audit scope will define the extent and limits of the audit: the parts of the ISMS that are to be covered, which activities, systems, processes, controls, locations and business units. The audit criteria shall include the requirements in ISO/IEC 27001 and any standards or requirements specific to the organization.

The audit shall be carried out by a qualified ISMS auditor or team of ISMS auditors, depending on the size and extent of the audit scope. A qualified ISMS auditor (or auditor team) will have the necessary knowledge, skills, competence and audit experience to undertake ISMS audits. If there is an audit team, then the team will be headed by a lead auditor and typically the lead auditor should have several more years of experience in ISMS auditing. Those that carry our ISMS audits should not have been involved in the development of the ISMS or have any current involvement in ISMS activities: in other words, they need to avoid any conflicts of interest and they need to remain impartial. It is only through impartiality of the audit that confidence and trust in the results can be assured. Internal ISMS audits and third-party certification audits therefore need to be objective, unbiased, free of prejudice, balanced and fair.

The general audit process, which is applicable to internal audits, is described in Chapter 10.

Internal audits are a key element in the monitoring and review process and are a useful management tool for identifying problems, risks and nonconformities related to the ISMS. The results and findings of the internal audit, documented in an audit report, are provided as input to the management review meetings (Section 8.6). The results also provide feedback to the senior management and interested parties.

8.6 Management Reviews of the ISMS

8.6.1 Management Review

Another aspect of the performance evaluation exercise is management review of the ISMS. Senior management shall have management reviews regularly at planned intervals. The purpose of these, like that of the overall aim of a performance evaluation, is to check the effectiveness, adequacy and suitability of the ISMS. These meetings need to review all relevant information, reports from audits and security assessments, results of measurements, risk assessments, feedback, changes affecting the organization, changes to requirements and consideration of other types of input that might have an impact on the effectiveness, adequacy and suitability of the ISMS.

These meetings should be attended by interested parties and all those that need to report into such meetings, because of their ISMS job function or because of an action that was placed on them, for example, by past meetings. Others might need to attend these meetings, not on a regular basis, where a specific agenda item requires their specific skills or knowledge about a relevant ISMS subject.

The agenda and minutes of these meetings shall be documented: this should be a record of discussions, decisions, actions and recommendations for ISMS improvements.

8.6.2 Input for the Management Review

Gathering information concerning changes should be part of an organization's monitoring, review and reporting process. The organization can get information, for example, from:

- Risk management records and reports.
- Incident handling records: These records can provide information on the organization's ability to resist against and respond to threats to its information systems. These records are a valuable source of information regarding the effectiveness of the organization's ISMS. Regular reviews of this information is essential to keeping up to date regarding the organization's responsiveness and preparedness for threats and attacks. Incident handling is key to the ISO/IEC 27001 ISMS process.
- ISMS measurements: Setting up metrics, a measurement programme and performance indicators is key to the ISO/IEC 27001 ISMS process. How effective, adequate and suitable are the leadership and support controls that management have in place? Is the training and awareness programme effective? How effective is the management of human resources? How effective is the incident-handling process? How well does IT backup perform? How good is the performance of the access control system to withstand attacks from external networks? These and many other questions can be asked to lead us to judge how effective our ISMS is at managing risks based on what we measure.

- Scorecards, gap analysis and benchmarking processes, tools and exercises: These methods, if used on a periodic basis across the organization and by an appropriate number of staff, can provide a valuable source of information. They give a relatively easy way of determining if security is being deployed and is working and whether target levels are being met.
- Internal audits and reviews and third-party certification audits: All forms of audit and review are important to determine if the ISMS is conforming to the requirements of ISO/IEC 27001 and whether the system of controls to manage the risks have been properly implemented, deployed and maintained; it is also necessary to review other audit results and to provide reports to management on the status of information security within the organization.
- Feedback from users, employees, customers, suppliers and other stakeholders: Often valuable information on the effectiveness of the organization can be obtained from staff. This should be the case because they are in daily contract with the system of risk controls in place. They are using the procedures that underpin the management and application of the information security measures in place. Also, customers and suppliers can sometimes provide useful information both positive and negative of the effectiveness of otherwise the organization's measures to protect information.

8.6.3 Output of the Management Review

8.6.3.1 Recommendations for Improvement

Gathering information and evaluating and analysing it with respect to the effectiveness, adequacy and suitability of the organization's ISMS implementation is a key exercise and one that proves the essential input into the management review. The next phase is that of deciding whether improvements need to be made, based on this input. What is decided are the recommendations and the actions to be taken for the main outputs of the management review.

Any changes that are recommended should be to improve the ISMS and change the ISMS for the better. Any changes to the ISMS to make improvements will have an influence and impact on the business. Any review recommendations need to be carefully considered; although it is aimed at improving the ISMS, it may have other impacts on other parts of the organization's systems and process.

Recommendations may be made requiring corrective actions to resolve existing problems and to make improvements. It is also the opportunity to take preemptive actions (to implement preventative controls) to prevent further problems from occurring: either old problems reoccurring or the anticipation of new problems occurring.

8.6.3.2 Action Plan

Any recommendations and actions to implement ISMS corrective and preventative actions to make ISMS improvements shall have full management commitment and support. An action plan needs to be produced with a prioritization of the work. After implementing the action plan, it needs to be confirmed that the new improvements are working effectively and are being deployed correctly in the working environment. Reworking, revising and updating of policies and procedures may need to be instituted. The improvement phase might need to confirm whether certain processes need to be improved and modified accordingly.

It is not only ISMS controls that may need to be improved but the ISMS processes themselves that may need to be improved and overhauled: the risk assessment and treatment process, the measurements programme and so on. This might entail changes to risk criteria, different methods of risk assessment deployed, new measures and performance indicators may need to define or existing ones changed or modified.

8.7 Awareness and Communications

Key to any action senior management might need to take, in this case with regard to ISMS improvements, is effective communication with all its managers and staff and all interested parties: awareness, training and deployment of procedures. Users and employees need to be aware of the changes to implement the improvements and how these relate to their own job function and the work responsibilities they have been assigned. Indeed, they need to practice deploying the procedures and processes they need to follow with regard to the management of risk in their areas of work.

Retraining may need to take place if the improvements involve totally new controls or upgrades to existing controls, with those staff that are directly impacted by these changes.

Communications need to be two-way. Management set the policy and strategy, make decisions and provide risk management resources. Users and staff should be reporting back to management incidents, giving feedback and providing information on potential risk situations and possible improvements to management; this is valuable input for the organization itself and to make informed decisions on how best to manage risks and the most appropriate course of action.

CHAPTER

9

Contents

9.1 Continual Improvement

9.2 Conformance and Nonconformance

9.3 Making Improvements

Improvements to the ISMS

The future is the shape of things to come.
—Accredited to H. G. Wells

9.1 Continual Improvement

9.1.1 Improvement

The process of improvement is referred to as a continual improvement in ISO/IEC 27001 and is a mandatory aspect of conformance with this standard. ISMS continual improvement is a top management commitment. A commitment that should be driven by the objectives set by management to have an ISMS that is effective, adequate and suitable for its business purposes and remains so throughout its operational life. This means to achieve continual improvement, the organization needs to consider the issues and requirements defined in ISO/IEC 27001 Clause 4 and to manage the risks and opportunities in ISO/IEC 27001 Clause 6, that is, to ensure that appropriate protection is in place to treat the risks, this could be information security controls and processes to detect, prevent, reduce and respond to undesired effects (e.g., from security incidents, breaches, disasters, compromises). It also implies that the ISMS produce the results and outputs that are expected.

The top management of the organization should always be looking for opportunities for improvement. One of the aspects of this is through the organization's performance evaluation activities as presented in ISO/IEC 27001 Clause 9: this includes monitoring and measuring the performance of information security processes and controls, internal audits and management reviews. It might be the case that the organization has received feedback from interested parties, such as its business partners or customers, on service delivery and availability. For example, business partners or customers might now be requesting additional information security requirements to be considered in addition to that already determined through a previous consideration of ISO/IEC 27001 Subclause 4.2. It might be that there are complaints or concerns from business partners about a recent spate of incidents in which the organization was involved, and they want some reassurance that the organization is doing something to improve its information security. For example, an organization experienced an increase in the number of hours of operational downtime and, subsequently, the inability to access information for business processes. This problem subsequently had an impact on business efficiency and customer dissatisfaction. After incident and problem analysis, it appeared that the downtime was due to the lack of availability of several business processes that rely on IT systems. The cause of the problem was due to a lack of regular maintenance, which was subsequently due to a lack of human resources and prioritization of jobs, that is, maintenance of critical business systems. Improvements were clearly needed in this case and action needed to be taken before the problem got out of hand and the business disruption got to a point where the impact was highly damaging to the organization's productivity, efficiency and reputation (in the case where the problem had an effect on external customers).

It could be that during a management review the results of the monitoring and measurement activities has highlighted several areas where there is opportunity to improve the performance of the ISMS. This could equally be the case with the results of an internal audit or from external third-party audits.

Therefore, maintaining and improving the effectiveness, suitability and adequacy of the ISMS are thus key to ensuring an appropriate level of protection of the organization's information and that its information policy and objectives are being changed to be compatible with its business direction and strategy.

9.1.2 Maintaining Effectiveness, Suitability and Adequacy

If an organization does nothing, that is, does not make ISMS improvements, then the protection of their information will, over a period, loses its effectiveness. This may happen over a period of months, weeks or even days, and the longer the problem is left unattended, the greater the impact. As discussed in Chapter 8, changes occur; some are in the organization's control and some are not. The degree of uncertainty regarding information security risks in today's fast-changing business environment means that organizations need to be proactive and adaptive to the risks brought about by such changes. So improvements will need to be made to ensure that the ISMS remains effective and in the light of such changes and subsequent risks that the organization faces. The information security risks with which organizations are being confronted are growing in frequency and complexity because of the uncertainty of the risks involved, and these might be caused by changes that are occurring monthly, weekly or daily, in some cases. Improvements are necessary to keep up-to-date with the changes to its risk profile. Some changes that happen are not classified as information security risks per se, but nevertheless they can have an impact on the organization and the effectiveness of its ISMS to protect its information. These changes could be driven by external factors, for example, changes to suppliers, environmental, infrastructure, conditions, legislation or the market conditions. The changes could also be driven by internal risk factors, for example, changes in business policy, strategy, staff resources, mergers, management decisions or changes to technology.

Continual improvement is an essential element in the ISMS life cycle through all the life-cycle stages of implementation, organizational and operational deployment, monitoring, measurements, reviews, maintenance, keeping up-to-date and improving the ISMS to achieve effective information security management.

9.1.3 Holistic Effectiveness

Continual improvement applies to the overall ISMS itself and does not just mean improvement of the effectiveness of the information security controls or processes. Continual improvement is also a recurring step-by-step process rather than a continuous process of improvement that refers to a nonstop process. This is important because the organization needs to achieve the right balance between identifying when improvements are necessary and when they should be implemented (based on available resources and the level of risk and business impact) as opposed to continually making improvements hour by hour, day by day. Clearly, if there are risks and impacts that have been identified that could compromise the organization and its ability to protect its information, then the organization needs to review its situation and take whatever is most appropriate to manage the risk. If there has been a major incident, business crisis or emergency, which could have an immediate impact on the well-being or survivability of the business, then action needs to be taken which may or may not involve ISMS improvements.

9.2 Conformance and Nonconformance

9.2.1 Nonconformity

An ISMS nonconformity is a nonfulfillment of a requirement of ISO/IEC 27001. This means a nonfulfillment of any of the "shall" requirements in ISO/IEC 27001 is a nonconformity. For example, there is a requirement that management shall review the organization's ISMS at planned intervals, and failure to do so is a nonconformity. All the requirements in ISO/IEC 27001 include the word "shall." The organization shall do risk assessments according to a planned schedule, it shall undertake a risk treatment process to select the appropriate options for treating the risks (managing the risks) and it shall determine an appropriate set of controls. For the exact wording of these statements, the reader is referred to the text in the standard itself. One of the tasks of an audit (internal or third party) is to check whether there are any nonconformities and, if nonconformities are found, then the organization should do whatever is necessary to remove such nonconformities.

9.2.2 Corrections

Once a nonconformity has been detected or reported, then a corrective action shall be implemented to eliminate the nonconformity and/or its cause. For example, if a current policy does not conform to a requirement of the standard, then this problem could be corrected by replacing this policy with one that does conform. Likewise, if a staff awareness programme is insufficient and so results in a nonconformity, then the programme should be replaced with an improved programme.

9.2.3 Corrective Actions and Root Causes

In the case of a corrective action, it is necessary to first determine the cause of the nonconformity. Determining this cause depends on the complexity of the problem. In some cases, a simple brainstorming exercise is all that is needed; in other cases, the root cause may be found through considering several different contributing factors (a chain of cause and effect) and a more detailed analysis might be needed. The root cause might be a people, process, method, management or technical problem, or a combination. The effectiveness of the corrective action will depend on whether the cause (or causes) has been correctly identified. The cause may have been accidental (a one-off mistake or error), or it might have been intentional or systematic, which could, if not dealt with, be a reoccurring problem. Therefore, if the cause is not properly identified or analysed, or a systematic cause treated as if it were a one-off problem or accident, then there is a likelihood of the problem reoccurring.

After determining the cause(s) of the nonconformity, then corrective action needs to be taken to eliminate the cause of the nonconformity. In general, then, the organization's response to a nonconformity should therefore be correction, cause analysis and corrective action or, alternatively, cause analysis, correction and corrective action. Merely correcting an existing nonconformity may not wholly resolve the problem, the root cause, and so a reoccurrence of the problem may happen.

9.2.4 Some Common Causes on Nonconformity

There is no definitive list of causes of nonconformities. The following are some of the common causes of nonconformities:

- Management failure to understand the internal and external issues and the requirements of interested parties that are relevant to the ISMS, which subsequently may have an impact on the delivering the expected outcomes of the ISMS in compliance with ISO/IEC 27001 Subclauses 4.1 and 4.2;
- Failure to adequately define the ISMS scope in compliance with ISO/IEC 27001 Subclause 4.3, especially a failure to properly address the interfaces and dependencies;
- Lack of an adequate or effective risk assessment and treatment process in compliance with ISO/IEC 27001 Clause 6;
- Management failure to commit to or to achieve continual improvement of the ISMS ISO/IEC 27001 Subclauses 5.2 and 6.1.1;
- Lack of resources to support the ISMS in compliance with ISO/IEC 27001 Clause 7;
- Lack of training and awareness in compliance with ISO/IEC 27001 Subclause 7.3;

- Failure to implement and control the processes necessary to effectively and adequately address the risks and opportunities in compliance with ISO/IEC 27001 Clause 6;
- Management failure to adequately address the requirements for performance evaluation in compliance with ISO/IEC 27001 Clause 9, including failing to comply with internal audit results, lack of commitment to management reviews and failure to determine a suitable programme of monitoring and measurements, all of which can fail to identify opportunities for ISMS improvements.

9.2.5 Case Study 1

An incident occurs regarding the operational use of security procedures to achieve the organization's security objectives. The incident is a failure to protect commercial information. A new inexperienced member of staff uses an "acceptable use" policy and procedure that was determined as necessary from a past risk assessment and risk treatment activity (ISO/IEC 27001 Subclauses 6.1.2 and 6.1.3). The inexperienced member of staff sends sensitive commercial information unprotected in an email based on their interpretation of the policy. The initial reaction was that the use of the procedure by the inexperienced member of staff was the cause but, after reviewing the procedures, it was found that the instructions given on this issue were unclear and badly written. So improvements should be made involving revision of the procedures and, in addition, providing better training to new recruits (ISO/IEC 27001 Clause 7.3).

9.2.6 Case Study 2

An auditor raises a nonconformity regarding management not providing the appropriate level of resources for the maintenance of the ISMS (ISO/IEC 27001 Subclause 7.1), in particular measurements for the evaluation of the performance of the ISMS. This results in a lack of appropriate input to the management review meetings (ISO/IEC 27001 Subclause 9.3 (c)). There is a lack of human resources to undertake performance measurements and to analyse and evaluate the ISMS effectiveness is the cause of the problem. Therefore, management needs to review this resourcing problem to improve the process of providing input on the performance of the ISMS to management review meetings.

9.2.7 Case Study 3

An incident occurs related to the organization's use of social media as part of their marketing campaign. In the risk assessment, the organization had identified, analysed and evaluated a number of risks (ISO/IEC 27001 Clause 6.1.2) relating to their use of social media for business purposes and they subsequently determined a number of controls (ISO/IEC 27001 Clause 6.1.3 (b)) to treat and manage these risks. Unfortunately, after 6 months, the organization has still only implemented a few of these controls leaving the organization wide open to the risks assessed. However, the organization had failed to implement the controls (ISO/IEC 27001 Subclause 8.1) determined in (ISO/IEC 27001 Subclause 6.1.3 (b)) to achieve the organization's security objectives (ISO/IEC 27001 Subclause 6.2). The lack of implementation of appropriate risk-based controls was the main cause of the incident, that is, lack of controls that need to be implemented to treat the identified risks.

9.3 Making Improvements

9.3.1 Planning and Implementing Improvements

Information on what improvements are needed, when improvements should or can be carried out and what resources are needed should all be documented as part of an improvement plan. The improvement process should ensure that:

- Improvements that are implemented are communicated to all interested parties; in particular, communicate to all those who need to know the details about the improvements (i.e., at a level of detail that is appropriate and relevant to them and their job function), be it regarding new security methods and measures, new or revised policies and procedures or other changes; communications might not be to just to internal staff but might need to be to external third parties, customers and other external parties that need to know about the changes (i.e., at a need-to-know level of detail that is appropriate and relevant to them in accordance with the contracts and SLAs in place) (see ISO/IEC 27001 Subclause 7.4).
- Action is taken to revise and update policies and procedures where necessary and appropriate as part of achieving continual improvement.
- Action is taken to revise processes where necessary and appropriate as part of achieving continual improvement (see ISO/IEC 27001 Subclause 8.1).

- Suitable awareness is undertaken to inform about the improvements, where this might involve changes or revisions to policies and procedures, other risk controls or processes and if it might also involve giving some staff training (see ISO/IEC 27001 Subclauses 7.2 and 7.3).
- There are adequate resources available needed to deploy and utilize the implemented improvements (see ISO/IEC 27001 Subclause 7.1).

9.3.2 Implementing Improvements to Management Leadership and Commitment

If top management fails to provide and demonstrate sufficient leadership, then this is a nonconformity. ISO/IEC 27001 Subclause 5.1 (a) through (h) covers a number of areas where leadership and commitment need to be demonstrated if conformance is to be claimed by the organization.

9.3.3 Implementing Improvements to Resources and Competence

Top management failing to provide sufficient resources can be a cause of a nonconformity (see ISO/IEC27001 Subclause 7.1). If the organization does not have a workforce with suitable and appropriate competence, then this is a nonconformity (see ISO/IEC 27001 Subclause 7.2).

9.3.4 Implementing Improvements to Awareness and Training

As mentioned above, lack of training is a common cause of a nonconformity and is always an area where there is opportunity for improvement. Does the organization need to improve the competence of some staff involved in ISMS-related activities through appropriate education and/or on-the-job training and experience? Does the staff need a refresher training as relevant to their work? Are the organization's awareness and training programme sufficiently comprehensive to cover the scope of its ISMS implementation?

9.3.5 Improvements to Processes

There are various situations where processes need to be improved, for example, where processes are not properly used or not followed, are difficult to use, not well designed or implemented, cause performance problems or result in nonconformities. It might be the case that the risk assessment process itself needs to be improved, as it is not producing consistent or valid results. Another example might be the process the organization uses to communicate with external parties needs to be improved. Also, the processes that organization deploys for outsourcing are not properly or inadequately controlled. The internal audit process may need to be reviewed and improved to ensure the impartially of audits are being maintained. There are many processes, both information security processes and business processes, where there could be opportunities for improvement.

9.3.6 Improvements to Policies and Procedures

There are various situations where policies and procedures need to be improved. Policies and procedures may be written in ways that are confusing or difficult to use, they may contain errors or mistakes or provide wrong information, they may be incomplete or not cover all possible situations or they might need updating to take into account new developments. All these factors can and do cause problems in achieving the organization's security objectives and the intended product of the ISMS. Is the information security policy appropriate to the organization's objectives and strategic mission? Does the policy cover the right things? Has the organization a suitable set of procedures necessary to implement the information security objectives? Are the procedures sufficiently clear and understandable to avoid any errors or mistakes when used? Does the scope of the procedures need to be extended and the content updated to cover new areas of development and change in business operations?

CHAPTER

10

Contents

10.1 Overview

10.2 International Certification

10.3 Certification and Accreditation

10.4 Standards Involved

10.5 ISMS Audits

Accredited ISMS Certification

A man who has committed a mistake and doesn't correct it is committing another mistake.

—Accredited to Confucius

10.1 Overview

Conformity assessment is a process that is used to determine whether a product, service, process, management system or person meets the requirements of a standard or other specification. Typical conformity assessment activities include accreditation, certification, inspection, validation, verification and testing. Conformity assessment may be carried out by first parties (e.g., self-declaration by a service provider or manufacturer), a second party (e.g. a declaration made by the client on the supplier) or a third party (e.g., a declaration by an independent body). Management systems such as ISO/IEC 27001 are involved in certification and accreditation activities.

This chapter explores the world of third-party ISMS certification: the basic ideas; which parties are involved and what their roles are; what the process involves; who does the audit and what qualifications are needed to be an auditor and, finally, the current international trend for ISMS certification.

10.2 International Certification

10.2.1 Global Take-Up

ISO/IEC 27001 ISMS certifications are carried out in over 70 countries and involve small, medium and large organizations from a diverse range of business sectors; some examples are:

- Telecoms and network services and suppliers;
- Financial institutions and insurance sector;
- Manufacturing sector;
- Utilities (electricity, gas, oil and water);
- Healthcare sector;
- Supply chains and logistics;
- Transportation sector;
- IT vendors, suppliers and services;
- Government departments and agencies;
- Retail sector;
- Entertainment, leisure and hospitality sector industry;
- Gambling and lottery sector;
- Sports sector;
- Research and development organizations;
- Not-for-profit organizations (e.g., charities);
- Professions (e.g., legal profession, accountancy);
- Academic sector.

The current trend in certifications is Asia, followed by Europe, then the Americas and, finally, Africa and the Middle East. The breakdown at the time of this writing is:

- Asia (including Australia and New Zealand), 36%;

- Europe, 33%;
- Americas, 20%;
- Africa and the Middle East, 11%.

The growth rate in all areas of the world is continually increasing, and predictions from the market are that in the near future a more balanced global spread will be achieved.

10.2.2 Motivation

This sector take-up and the diversity within sectors demonstrate not only the importance of information security in every business sector but also the wide applicability of the ISO/IEC 27001 to meet the demands of business for a "common language" for addressing information security issues. The motivational benefits for going for a third-party ISMS audit varies, for example:

- Demonstrating your ISMS meets the requirements of the standard and providing stakeholders and interested parties with a degree of creditability, trust and assurance in the ISMS, showing the ISMS is "fit for purpose" and managing the risks effectively;
- Compliance with legislation and regulations (e.g., cybersecurity, cyber-resilience, data privacy and protection, governance and computer misuse and hacking);
- Requirements in customer and supplier chain contracts (there are more tenders and RFPs that are now including ISO 27001 compliance clauses);
- Insurance reasons (there are instances where insurance premiums have been lowered as organizations can demonstrate they have an effective risk management system in place);
- Market competition (demonstrating ISMS excellence in a particular sector).

10.2.3 Costs and Resources

The cost to the end-user organization of implementing an ISMS and having the ISMS certified breaks down into two components: external costs and resources (initial certification audit, follow-up surveillance audits and recertification audits) and the cost of internal work (preparing for certification, ongoing surveillance audits and maintenance and operational costs). How these costs are calculated varies and is dependent on several factors, for example, the size of the organization and the complexity of its ISMS scope.

With regard to the internal costs and resources, this includes:

- Organizational and operational processes;
- Implementing risk management processes and associated activities;
- Implementing and deploying ISMS determined controls;
- Performance evaluation and continual improvement activities;
- Producing documentation;
- Leadership support, awareness and training;
- Allocating resources for specific security functions (e.g., a security officer or manager, firewall administration or physical security manager).

This might seem an expensive exercise, but some of these items should be everyday expenses for most organizations that are protecting its information assets.

Also, depending on the size and system's complexity, these costs maybe not be large at all. In effect, if security is integrated across the organization, as it should be, with other parts of the business, then some of these costs will be absorbed and shared by other business functions and operations. Information security is an important aspect of any business, and it is rare to find a company where some aspect of information security is not required or implemented. Information security should be treated as an investment if an organization wants to protect one of its most valuable assets, its information. This means resources should be part of the business budget whether the organization goes for certification. It is a commercial balance and trade-off between implementing just the appropriate level of security and going over the top with unnecessary expenditure. Therefore, it is important to carry out an information security risk management process so that management can make the right decisions on what to spend money (see ISO/IEC 27001 Clause 6).

Hence, we must not lose sight of the fact that the business needs to survive in today's dynamic and competitive markets and so the more accurate its analysis of the risks the company faces, the better informed is management to make the right decisions to do the best for the company and how to best invest in information security.

For small businesses, the costs of implementing an ISMS and for ISMS certification will be a scaled-down version of that for a large organization. The size of the small business means the costs should be less. Also, the complexity of operations is generally relatively smaller than that of larger organizations.

10.3 Certification and Accreditation

10.3.1 ISO Standards and Definitions

The ISO has a committee CASCO that is responsible for conformity assessment in ISO. The development of policy and standards related to conformity assessment is undertaken by CASCO on behalf of the ISO. The CASCO standards provide the definitions and concepts for conformity assessment. These standards contain information about the accreditation and certification process and specify accreditation requirements.

CASCO has developed a toolbox of standards for use in conformity assessment, including the ISO/IEC 17000 series of standards. For example:

- ISO/IEC 17000 Conformity Assessment—Vocabulary and General Principles;
- ISO/IEC 17021-1 Conformity Assessment—Requirements for Bodies Providing Audit and Certification of Management Systems—Part 1: Requirements;
- ISO/IEC 17023 Conformity Assessment—Guidelines for Determining the Duration of Management System Certification Audits;
- ISO/IEC 17024 Conformity Assessment—General Requirements for Bodies Operating Certification of Persons;
- ISO/IEC 17028 Conformity Assessment—Guidelines and Examples of a Certification Scheme for Services.

It is important to note that neither ISO nor CASCO perform conformity assessment activities, such as certification. It is also important to note that the use of the ISO logo in relation to certification is forbidden. Certifications are carried out by bodies external to the ISO and not by the ISO.

The following are ISO definitions:

- *Certification:* Third-party attestation that a set of specified requirements have been fulfilled as demonstrated to the third party. This can be related to products, processes, systems or persons, and, in the case of the ISMS, this is a set of management system requirements specified in ISO/IEC 27001. The third party is referred to as a conformity assessment body, and, in the case of a management system such as ISO/IEC 27001, these are commonly referred to as a certification body.

- *Certification audit:* Audit carried out by an auditing organization independent of the client and the parties that rely on certification, for the purpose of certifying the client's management system.
- *Accreditation:* Third-party attestation that a conformity assessment body has formally demonstrated its competence to carry out specific conformity assessment tasks.

10.3.2 Interested Parties Involved in Certification

In addition to the ISO and CASCO, there are various other stakeholders interested in conformity assessment and those that provide accreditation or certification services. This includes:

- *Accreditation body (AB):* An independent authoritative body that undertakes accreditation according to the CASCO definition above. An AB is generally derived and established from government as a national AB. All countries have national ABs.
- *Certification body (CB):* An independent body that provides certification services to end-users (organizations that have ISMS implementations).

Each of these interested parties plays a different role in the process and has different responsibilities: the AB's role is accreditation, and the CB's role is certification.

The International Accreditation Forum (IAF; www.iaf.nu) is the global association of accreditation bodies and other bodies interested in conformity assessment. The main function of the IAF is to develop a single worldwide programme of conformity assessment aimed at reducing risk for businesses and providing assurance that accredited certificates and validation and verification statements may be relied upon. There are over 90 ABs that are members of the IAF; the list of members can be found at www.iaf.nu.

Each national AB will have a list of accredited organizations, that is, CBs that offer certification services. Organizations can search for a CB by contacting a national AB or, in most cases, searching their website for a list. Examples of ABs are ANAB in the United States, cofrac in France, CNAS in China, DAkkS in Germany, EMA and MAAC in Mexico, ISMS-AC and JAB in Japan, SCC in Canada, SWEDAC in Sweden and UKAS in the United Kingdom.

10.3.3 Accreditation

Accreditation is a formal process carried out by a national AB. The AB provides formal, third-party recognition of the competence of the CB to perform specific certification and auditing tasks and deliver certification services to the end-user. For the end-user organization (CB customer), it provides a means to identify a proven, competent CB so that the selection is an informed choice. It also means the CB can demonstrate to its customers that it has been successful at conforming to the requirements specified in international accreditation standards for management systems.

The ISMS certification process uses the same model as other management systems such ISO 9001 for quality management, ISO 14001 for environment management and ISO 220002 for food safety management.

In assessing the adequacy and competence of the CB to carry out ISMS certifications in accordance with ISO 17021 and ISO/IEC 27006-1, the AB will:

- Carrying out CB head office visits to check that the CB has the necessary systems, processes and procedures in place, and these are functioning correctly to undertake such ISMS audits.
- Assess whether the CB has an adequate number of qualified auditors to carry out ISMS certifications.
- Carry ISMS on-site for witnessed audits to assess the competency of the auditors working in practice with the CB's customers.

After being accredited, the CB will be awarded a certificate of accreditation, which shows their competence in being able to offer ISMS certification services to their clients. This accreditation lasts for 3 years, during which their competence and performance will be continually monitored and assessed by the AB through regular witnessed audits. After 3 years, the CB can be reaccredited by the AB, and this can extend their certificate of accreditation for another 3 years.

A CB that offers accredited certification services needs to be accredited by an AB. Currently, there are many CBs that can provide accredited ISMS certification services.

10.3.4 Certification

The CB carries out certification audits on their client's ISMS. In Section 10.5, we describe the audit process, which is a two-stage process.

The CB issues certificate to its client based on a successful audit being carried out. This certificate lasts for 3 years. During this 3-year period, the

CB will carry out regular surveillance audits every 6 to 12 months. At the end of the 3 years, the certificate expires. If the client wishes to continue to have a valid certificate, then, before the expiry date, the client needs to apply to the CB for a recertification audit. The client will be reissued with a certificate following a successful recertification audit.

10.3.5 Selecting an Accredited CB

The following are important and valued advice for end-user organizations (ISMS owners) to consider when selecting an accredited CB:

- Evaluate 2 to 3 CBs before selecting one.
- Enquire whether the CB uses the relevant CASCO standard (see Section 10.4).
- Find out whether the CB has been accredited by an AB. Remember accreditation gives independent endorsement of the competence of the CB.

It is important to verify that a CB has been accredited by a recognized national AB (see the list of ABs on https://iaf.nu/en/accreditation-bodies/), as this provides guarantees as to the competence of a CB to provide certification services. Accreditation also provides recognition that the delivered attestation will be recognized by other CBs, by ABs, governments and regulatory bodies, nationally, regionally and internationally. It is important to note that accreditation applies to a specific scheme or standard. For example, a CB may be accredited to provide certification services for ISO 9001 but not for ISO/IEC 27001. However, the CB may be accredited to provide services for ISO 9001, ISO/IEC 27001 and several other management systems standards. It is therefore important to research and check whether the CB is accredited. The current list of accredited CBs is shown at https://www.iafcertsearch.org/search/certification-bodies.

10.4 Standards Involved

The main standards involved in the accreditation and certification process are the following.

10.4.1 Accreditation

The following are the current standards used by the AB to assess and accredit the CB to provide ISMS certification services:

- ISO 17021-1—Conformity Assessment—Requirements for Bodies Providing Audit and Certification of Management Systems;
- ISO/IEC 27006-1—International Accreditation Guidelines for the Accreditation of Bodies Operating Certification—Registration of Information Security Management Systems.

10.4.2 Certification

The following standard is used by the CB to audit and certify the customer's ISMS:

- ISO/IEC 27001—Information Security Management Systems—Requirements.

10.4.3 Supporting Auditing Guidance (Not Mandatory)

- ISO 19011—Guidelines for Auditing Management Systems;
- ISO/IEC 27007—Guidelines for Information Security Management Systems Auditing;
- ISO/IEC 27008—Guidelines for Auditors on ISMS Controls.

10.4.4 End-User Organizations (ISMS Owners)

Organizations may use all the other standards in the ISO/IEC 27000 ISMS series as supporting guidance and advice on the implementation of ISO/IEC 27001. This includes ISO/IEC 27003—Information Security Management System Guidance, ISO/IEC 27004—Information Security Management—Monitoring, Measurement, Analysis and Evaluation and 27005—Information Security Risk Management.

Note that these supporting standards are not mandatory for certification audits. They do not need to be used by the end-user organization and should not be used by auditors as mandatory conformance standards. It is entirely up to the end-user organization (ISMS owner) whether they need this guidance.

10.5 ISMS Audits

10.5.1 Certification Scope

The scope of the ISMS (as detailed in ISO/IEC 27001, Chapter 4.3) is defined by the organization. The role of the CB to check that the scope of certification (which could be different) is consistent with its business and that it does not exclude anything of their operation, which could affect the organization's capability, and/or responsibility, to provide information security that meets the security requirements determined by risk assessment and applicable regulatory requirements.

The CB should therefore ensure that the organization's information security risk assessment properly reflects its activities and extends to the boundaries of its ISMS activities, taking account of any interfaces and dependencies as defined in ISO/IEC 27001 (Chapter 4.3), and it should confirm that this is reflected in the organization's Statement of Applicability (SoA); see ISO/IEC 27001 Chapter 6.1.3.d). If the ISMS interfaces with services or activities that are not completely within its scope, the risks associated with these interfaces should nevertheless be included in the organization's information security risk assessment.

10.5.2 Audit Process

The certification of an organization's ISMS against the standard ISO/IEC 27001 is generally a two-stage audit process as detailed in ISO/IEC 26006-1 and ISO 19011.

10.5.2.1 Audit Stage 1

During this stage, the CB carries out an initial assessment to gain knowledge and understanding of the auditee's ISMS and its business. The CB should examine the documentation covering the design and implementation of the ISMS. This will include consideration of the definition of the ISMS scope and its ISMS policy, the risk assessment report and risk treatment plan, the SoA and other core elements of ISMS. This review enables the auditors to get an understanding and appreciation of the organization's ISMS scope, policy and design. This then will provide a focus for planning the subsequent parts of the audit. It gives an indication of the state of preparedness of the organization for the audit. Once this review is complete and an audit report is produced, the CB is able to decide whether to proceed with Stage 2 of the audit. Before proceeding to the next stage, the CB needs to bring together an audit team with the necessary competence to address the auditing of the organization's specific ISMS scope.

10.5.2.2 Audit Stage 2

During this stage, the CB audit team visits the site(s) of the organization to carry out an audit of the ISMS. The objective of this stage is to confirm that the organization complies with the requirements of the ISO/IEC 27001 standard and that adheres to its own policies and business objectives.

The audit team should clearly focus on the main areas of the standard's mandatory requirements:

- The ISMS in the organizational context (ISO/IEC 27001 Clause 4);
- Commitment, leadership and support of the organization's top management (ISO/IEC 27001 Clauses 5 and 7);
- The risk assessment and risk treatment processes (ISO/IEC 27001 Clause 6);
- The resulting ISMS design and implementation (ISO/IEC 27001 Clause 8);
- The performance evaluation process (ISO/IEC 27001 Clause 9) and the continual improvement (ISO/IEC 27001 Clause 10). The audit team also needs to consider specifics, for example, what the organization has put in place for monitoring and measuring ISMS performance and effectiveness, as well as reporting and reviewing against its business objectives and targets.

The assessment will include checking:

- That what has been planned and implemented is in operational use to ensure that security;
- That management reviews take place;
- That there is an internal ISMS audit process in place;
- That management responsibility has been defined for the implementation and deployment of its information security policy.

The audit team also needs to consider specifics regarding the risk management process: the risk assessment methods used, the risk criteria, the process of determining controls, the SoA and the justification statements in the SoA.

An important aspect of the audit is the presentation and examination of objective evidence to demonstrate the audit trail links between the information security policy, the results of the risk assessments, security objectives and targets, the responsibilities for information security, the system of controls and procedures and what has been put in place for performance monitoring and security reviews.

10.5.3 Nonconformities

In the case of ISO/IEC 27001, a nonconformity is the absence of or the failure to implement and maintain one or more required ISMS elements or a situation, which would, based on objective evidence, raise significant doubt as to the capability of the ISMS to achieve the security policy and objectives of the organization.

The CB can define grades of deficiency and areas for improvement (e.g., major or minor nonconformities and observations). If such grades of deficiency are used, however they may be defined by the CB, all types of nonconformity should be dealt with as specified in ISO 17021-1 and ISO/IEC 27006-1.

10.5.4 Audit Report

The audit team is required to provide to the CB a report of its findings and recommendations as to the conformity of the organization's ISMS with all the requirements of the ISO/IEC 27001 standard. Any nonconformity to be discharged in order to comply with all the requirements of the ISO/IEC 27001 standard should be promptly brought to the organization's attention.

The CB shall ask the organization to comment on the report and to describe the specific corrective actions it will take or plans it will take within a defined time to remedy any nonconformity. The closure of such follow-up actions may need a full or partial reassessment of the ISMS or a written declaration to be confirmed during surveillance may be considered adequate.

This report should include the identification of ISMS elements audited, the assessed scope of the ISMS being audited, comments on the conformity of the organization's ISMS with the requirements of the standard with a clear statement of any nonconformity and, where applicable, any useful comparison with the results of previous assessments of the organization.

The decision whether to award a certificate regarding an organization's ISMS shall be made by the CB on the basis of the information gathered during the audit, the audit report and any other relevant information. Those who review the report and make the decision to certify shall not have been those that have participated in the ISMS audit.

The certificate awarded to the organization carries the logo of the CB as well as the logo of the accreditation body that has accredited the CB.

10.5.5 Surveillance Audits

The CB will carry out periodic surveillance audits at sufficiently close intervals to verify that the organization's certified ISMS continues to comply with the requirements of the ISO/IEC 27001 standard and the ISMS remains effective regarding achieving the objectives of the organization's information security policy. This audit should also examine the action taken by the organization on nonconformities identified during the last audit.

If nonconformities are found during surveillance audits, then these shall be effectively corrected within a time agreed by the CB. If correction of any nonconformity is not made within this agreed time, then the certification shall be suspended or even withdrawn. The time given to carry out corrective action should be consistent with the severity of the nonconformity.

According to the ISMS standard ISO/IEC 27001, the organization needs to conduct internal ISMS audits at planned intervals to determine whether the ISMS system of controls, processes and procedures conforms to the requirements of the standard. At the same time, the audit should take into account the relevant legislation or regulations requirements and that the ISMS has been effectively implemented, is effectively maintained and performs as expected. An audit programme shall be planned to take into consideration the status and importance of the ISMS processes and areas to be audited, as well as the results of previous audits.

10.5.6 Recertification

Recertification is normally carried out every 3 years and its purposes are to verify the overall continuing conformity of the organization's ISMS to the requirements of the ISO/IEC 27001 standard and that the ISMS has been properly implemented and maintained.

Recertification and surveillance audit programmes should normally include:

- Verification that the approved ISMS continues to be implemented.
- Consideration of the implications of changes to the ISMS of controls initiated as a result of changes in the organization's operation and ensuring the overall effectiveness of the ISMS in its entirety in the light of changes in operations. This includes changes to the ISMS documented system.
- Confirmation of continued compliance with the requirements of ISO/IEC 27001 and a demonstration of the commitment to maintain the effectiveness of the ISMS and the effective interaction between all elements of the ISMS.
- Aspects of system maintenance, which are the internal ISMS audit, internal security review, management review, and preventive and corrective action.

The ISMS processes, as specified in the ISO/IEC 27001 standard, provide a management framework for an organization to realize a programme of continual improvement. This should provide the organization with effective ISMS implementation and deployment for the common good and governance of its business and also matches the aims of the recertification and surveillance audit activities mentioned above.

10.5.7 Audit Trails

An important aspect of a certification audit is the collection of objective evidence. In gathering such evidence, an auditor commonly looks for a sample of audit trails. For example, the organization should have carried out a risk assessment, risk treatment, and selection of controls exercise leading to an SoA. The auditor should be able to backtrack this process looking for evidence that the organization has carried out the process correctly, for example:

- The SoA shows a specific control set has been selected for the implementation question: What was the decision-making behind this selection?
- Tracing backwards, the auditor might then want to discuss this decision-making process in relation to the risk treatment options selected.
- Tracing back further, the auditor might then ask: What risks were identified in the risk report that led to the risk treatment option of reducing the risks?
- Tracing back, the auditor might want to look at the risk assessment report and discuss the control set that has been selected commensurate with the assets, threats and vulnerability to which they relate and the associated risks that have been identified.

The auditor will want to check by gathering answers and evidence that there is a correspondence from the controls in the SoA back to the risk assessment that has been carried out. Other examples include checking on the incident-handling process that is in place from:

- Detection, reporting and recording the incident;
- The analysis of the incident;
- The corrective actions taken to resolve and recover from the incident;
- The closure of the incident action;

- Follow-up actions and activities to ensure preventative measures are deployed to avoid reoccurrence of the incident.

Again, the auditor could build up an audit trail of objective evidence that the correct procedures and deployment incident handling controls are in place to manage the risks of incident past, present, and future. There are many examples in the implementation and deployment of the controls from ISO/IEC 27001 Annex A (i.e., the control statements from ISO/IEC 27002). These might include establishing audit trails relating to:

- Access control measures and procedures;
- Backups;
- Business continuity plans and testing;
- Email and internet policies;
- Legal and regulatory controls;
- Third-party contracts, SLAs and service delivery management;
- Physical security measures.

Like the two examples given earlier, the auditor should be looking to build up audit trails of objective evidence that the correct procedures and deployment of controls are in place to manage the risks of the business. The auditor will also be looking at the ongoing management of these processes and controls looking for indications and evidence that the organization is monitoring and reviewing its systems to check how effective its security is and what measures are planned to improve in areas where the security is not so effective.

10.5.8 Competence

Confidence and reliance in the certification audit process are dependent on several things: the general competence of personnel involved in each of the certification functions, and the more specific competence of the auditors conducting ISMS audits. The certification functions include conducting client application reviews, auditing, being a lead auditor, and reviewing audit reports and making certification decisions.

The accreditation requires that the CB has a process in place that defines the necessary competence requirements of its personnel in the certification functions; this includes technological, legal and regulatory knowledge relevant to the client's ISMS being audited.

The CB shall also have a process and criteria in place for verifying the competence of personnel: their experience, training and knowledge. General and specific experience, knowledge and training related to ISMS certifications include:

- Management systems and auditing;
- Principles of auditing and the audit process (ISO 19011);
- Management systems principles, philosophy, processes, ISMS-specific;
- General knowledge of information security;
- ISMS standard ISO/IEC 27001;
- Specific processes such as risk assessment and risk treatment;
- ISMS monitoring, measurement, analysis and evaluation;
- Information security controls relevant to implementing the ISMS standard ISO/IEC 27001;
- Technical knowledge of the activity to be audited.

The CB needs to have an effective process in selecting auditors with the necessary knowledge, skill and experience to ensure that the audits are carried out effectively and uniformly. The general criteria for auditor competence are based on a demonstration of the ability to apply a range of knowledge and skills gained from the formal education, IT and ISMS work experience and ISMS auditor training and auditing experience.

Auditors are expected to develop, maintain and improve their competence through a programme of Continual Professional Development (CPD) and by participating regularly in ISMS audits.

CHAPTER

11

Contents

11.1 The ISMS: A Living System

11.2 Processes Cycles and Sequences of Activity

11.3 ISMS the Business Enabler

Coda

It is in changing, that things find purpose.
—Accredited to Heraclitus

11.1 The ISMS: A Living System

The ISMS should not be treated as a static system. It needs to be treated as a living, dynamic system that will change over time to be kept up-to-date with the changes in the business environment in which it operates. If the ISMS is treated as a static system, then after a period of time, it will eventually become ineffective at protecting the organization's information assets: it will provide inadequate quality of protection, unsuitable to meet the requirements, needs, purpose and objectives of the organization and interested parties. Such an ISMS will devalue the organization's investment in the design, development and implementation of its ISMS.

Like all management systems, the direction comes from the top and this applies equally to ISO/IEC 27001 and the implementation of an ISMS; top management shall be the driver, policymaker, providing leadership, commitment, governance and support on behalf of the organization.

The only way the ISMS can maintain the three objective measures of effectiveness, adequacy and suitability is if the ISMS is treated as a living, dynamic and continually improved system (Figure 11.1).

This means that there are several critical processes and life cycles that need to be reviewed, refreshed and continually improved. These have been discussed in the preceding chapters. As a conclusion to this book, we shall below briefly summarize some of these processes.

11.2 Processes Cycles and Sequences of Activity

11.2.1 Information Life Cycle

Information is at risk at all stages of the information life cycle (see Figure 11.2). Information might be: on paper or in electronic form, spoken or oral, processed manually or using technology, single format, mixed or multimedia format, in small to large volumes (mega, giga, terra, peta and so on), real-time, nonreal-time/batched or streamed.

Throughout the life cycle, the risk to preserving the confidentiality, integrity and availability of information is essential:

- Confidentiality: protected against unauthorized access or exposure, leakage;
- Integrity: protected against any unauthorized or accidental changes of modifications to ensure its accuracy and completeness, provenance;

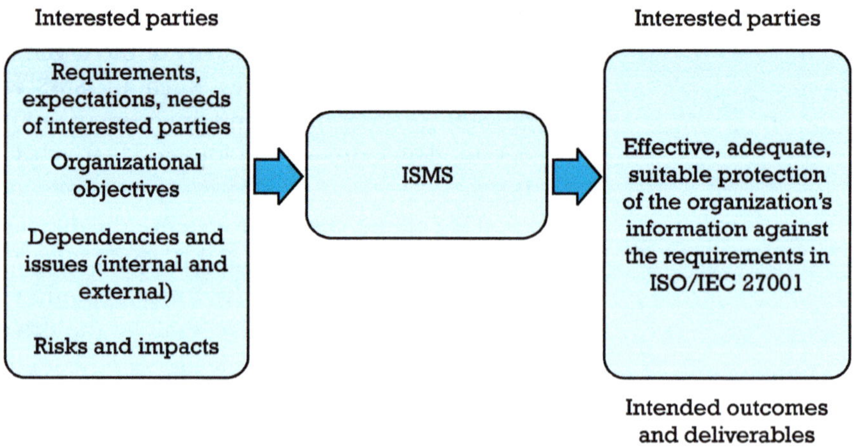

Figure 11.1 ISMS requirements and deliverables.

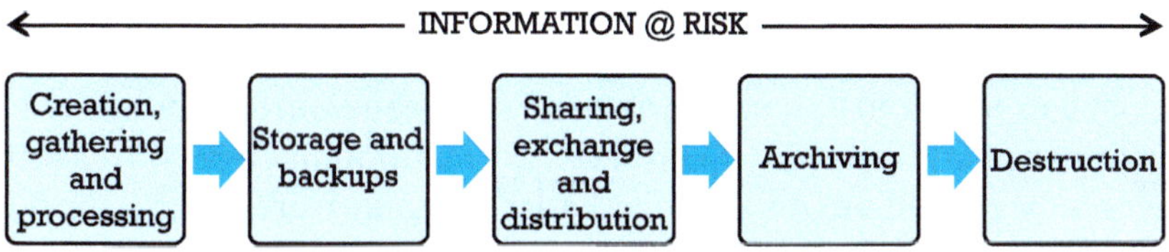

Figure 11.2 Information life cycle.

- Availability: protected to ensure that it is available to all those authorized to access the information and such information has not been denied to those authorized people.

11.2.2 Risk Management

The purpose of risk management is to identify, analyse and evaluate the risks to information and then to decide how to mitigate these risks through the adoption of a suitable selection of risk treatment options and the implementation of appropriate controls. These two processes are the basis for a sound strategic approach for information security and requisite for ensuring effective protection of the organization's information.

Implementing risk treatment options and information security controls follows on from the risk management exercise. Implementation needs to be carried out properly to ensure the treatment options and controls will be effective to mitigate the risks involved.

Risks change as time goes by and the mitigation of an organization's risks will likely become ineffectual over time and provide progressively less protection. The key to long-term sustainable protection is to make sure the organization assesses its risks on a regular basis and to upgrade and improve its treatment and protection accordingly. Risk management performed on a regular basis provides the capability for the organization's ISMS to respond to change, to be adaptive to change and to survive the risks, threats and attacks to which their information is subjective.

11.2.3 ISMS Performance and Improvements

Change is an inevitable aspect of every part of life and business. The ISMS is founded on a set of requirements, needs and expectations of the organization, all of which are likely to change over time. Managing this change and adapting to this change will ensure the ISMS remains effective. This includes consideration of the changes to risks and the security incidents that test the resilience of the ISMS to remain protected. This leads us to the topic of the performance of the ISMS and checking the ISMS is functioning and delivering protection as expected.

One aspect of this is monitoring and taking measurements of the ISMS performance and capability to protect the organization's assets. This activity needs to be carried out on a regular basis taking measurements against a set of predefined metrics and targets. The results need to be reviewed and assessed and a report produced for management and interested parties.

Another aspect is to carry out internal information security audits to check the ISMS maintains its conformance with the requirements of ISO/IEC 27001. An audit report should be produced and made available to management, and any follow-up actions to correct any nonconformities need to be dealt with.

Finally, a management review needs to be undertaken that will consider:

- Any changes in business and interested party requirements, expectations relating to the ISMS;
- Reports relating to risk management activities, performance (monitoring and measurement) activities, incident handling, internal ISMS audits;
- Feedback from interested parties, customers and clients, internal employees and management.

The management needs to make recommendations based on the above review to carry out the necessary action to make improvements to the ISMS to keep it robust, resilient and fit for purpose against risks and incidents that the organization may face at any time.

11.3 ISMS the Business Enabler

ISO/IEC 27001 is first and foremost a management system standard for managing information security. Therefore, the success of the ISMS is centred on top management commitment, leadership and policy. Hence, the reason why all the requirements of ISO/IEC 27001 are top management and organization-centred. Also, ISO/IEC 27001 requires stakeholder/interested party engagement including investors, employees, customers, business partners and regulatory bodies.

ISMS risk management and performance evaluation are also key topics for achieving effective, suitable and appropriate protection. If the ISMS is implemented, used, monitored and reviewed and continually improved to the deliver the intended outcomes against the requirements of ISO/IEC 27001, then the ISMS provides value as a business enabler helping the

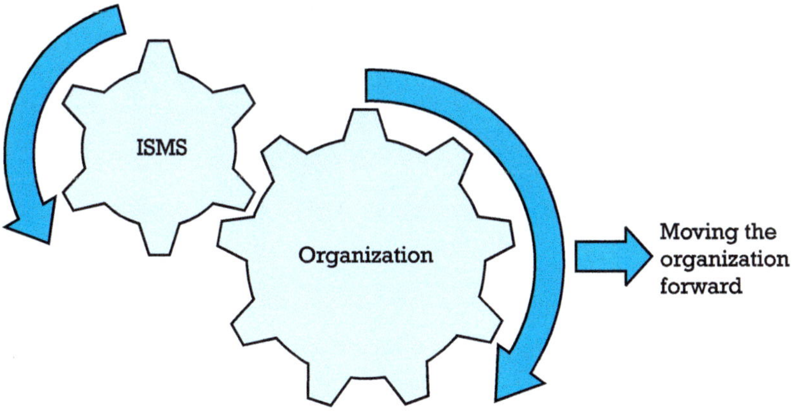

Figure 11.3 ISMS drives business forward

organization to drive forward knowing that its information is effectively, adequately and suitably protected (see Figure 11.3).

Maintaining a viable business in a world that is competitive, dominated by rapidly changing digital technology, intimidated by frequent, costly and damaging cyber incidents, is a challenge. Survival in this world requires being adaptive and responsive. ISO/IEC 27001 is a standard that is designed to be adaptive and responsive to the challenges of cyber risks and changes to the business environment and to continually improve its capability to protect its information assets.

Bibliography

It is important that readers and users refer to the latest (dated) versions of the ISO/IEC and ISO standards listed below. The latest versions are given in https://www.iso.org/committee/45306/x/catalogue/p/1/u/0/w/0/d/0.

[1] ISO/IEC 27000—Overview and Vocabulary.

[2] ISO/IEC 27001—Information Security Management Systems—Requirements.

[3] ISO/IEC 27002—Information Security Controls.

[4] ISO/IEC 27003—Information Security Management System—Guidance.

[5] ISO/IEC 27004—Information Security Management—Monitoring, Measurement, Analysis and Evaluation.

[6] ISO/IEC 27005—Guidance on Managing Information Security Risk Management.

[7] ISO/IEC 27006-1—Requirements for Bodies Providing Audit and Certification of Information Security Management Systems.

[8] ISO/IEC 27007—Guidelines for Information Security Management Systems Auditing.

[9] ISO/IEC TR 27008—Guidelines for the Assessment of Information Security Controls.

[10] ISO/IEC 27010—Information Security Management for Inter-Sector and Inter-Organisational Communications.

[11] ITU-T X.1051|ISO/IEC 27011 Information Security Controls Based on ISO/IEC 27002 for Telecommunications Organizations.

[12] ISO/IEC 27013 Guidelines on the Integrated Implementation of ISO/IEC 27001 and ISO/IEC 20000-1.

[13] ITU-T X.1054|ISO/IEC 27014—Governance of Information Security.

[14] ISO/IEC TR 27016—Information Security Management—Organisational Economics.

[15] ITU-T X.1631 | ISO/IEC 27017—Information Security Controls Based on ISO/IEC 27002 for Cloud Services.

[16] ISO/IEC 27019—Information Security Controls for the Energy Utility Industry.

[17] ISO/IEC 27021—Competence Requirements for Information Security Management Systems Professionals.

[18] ISO/IEC TS 27022—Guidance on Information Security Management System Processes.

[19] ISO/IEC TR 27024—Government and Regulatory Use of ISO/IEC 27001, ISO/IEC 27002 and Other Information for Security Standards.

[20] ISO/IEC 27028—Guidance on ISO/IEC 27002 Attributes.

[21] ISO/IEC TS 27100—Cybersecurity—Overview and Concepts.

[22] ISO/IEC 27102—Guidelines for Cyber-Insurance.

[23] ISO/IEC TR 27103—Guidance on Using ISO and IEC in a Cybersecurity Framework.

[24] ISO/IEC TS 27110—Cybersecurity Framework Development Guidelines.

[25] ISO/IEC 27031—Cybersecurity—Information and Communication Technology Readiness for Business Continuity.

[26] ISO/IEC 27032—Cybersecurity—Guidelines for Internet Security.

[27] ISO/IEC 27033-1—Security Techniques—Network Security—Part 1: Overview and Concepts.

[28] ISO/IEC 27033-2—Security Techniques—Network Security—Part 2: Guidelines for the Design and Implementation of Network Security.

[29] ISO/IEC 27033-3—Network Security—Part 3: Reference Networking Scenarios—Threats, Design Techniques and Control Issues.

[30] ISO/IEC 27033-4—Network Security—Part 4: Securing Communications Between Networks Using Security Gateways.

[31] ISO/IEC 27033-5—Network Security—Part 5: Securing Communications Across Networks Using Virtual Private Networks (VPNs).

[32] ISO/IEC 27033-6—Security Techniques—Network Security—Part 6: Securing Wireless IP Network Access.

[33] ISO/IEC 27033-7—Network Security—Part 7: Guidelines for Network Virtualization Security.

[34] ISO/IEC 27035-1 Information Security Incident Management—Part 1: Principles and Process.

[35] ISO/IEC 27035-2—Information Security Incident Management—Part 2: Guidelines to Plan and Prepare for Incident Response.

[36] ISO/IEC 27035-3—Information Security Incident Management—Part 3: Guidelines for ICT Incident Response Operations.

[37] ISO/IEC 27035-4—Information Security Incident Management—Part 4: Coordination.

[38] ISO/IEC 27036-1—Cybersecurity—Supplier Relationships—Part 1: Overview and Concepts.

[39] ISO/IEC 27036-2—Cybersecurity—Supplier Relationships—Part 2: Requirements.

[40] ISO/IEC 27036-3—Cybersecurity—Supplier Relationships—Part 3: Guidelines for Hardware, Software and Service Supply Chain Security.

[41] ISO/IEC 27036-4—Information Security for Supplier Relationships—Part 4: Guidelines for Security of Cloud Services.

[42] ISO/IEC 27037—Guidelines for the Identification, Collection, Acquisition and Preservation of Digital Evidence.

[43] ISO/IEC 27038—Specification for Digital Redaction.

[44] ISO/IEC 27039—Selection, Deployment and Operation of Intrusion Detection and Prevention Systems (IDPS).

[45] ISO/IEC 27040—Storage Security.

[46] ISO/IEC 27041—Guidance on Assuring Suitability and Adequacy of Incident Investigative Method.

[47] ISO/IEC 27042—Guidelines for the Analysis and Interpretation of Digital Evidence.

[48] ISO/IEC 27043—Incident Investigation Principles and Processes.

[49] ISO/IEC 27050-1—Electronic Discovery—Part 1: Overview and Concepts.

[50] ISO/IEC 27050-2—Electronic Discovery—Part 2: Guidance for Governance and Management of Electronic Discovery.

[51] ISO/IEC 27050-3—Electronic Discovery—Part 3: Code of Practice for Electronic Discovery.

[52] ISO/IEC 27050-4—Electronic Discovery—Part 4: Technical Readiness.

[53] ISO/IEC 19086-4 Cloud Computing—Service Level Agreement (SLA) Framework—Part 4: Components of Security and of Protection of PII.

[54] ISO/IEC 20547-4—Big Data Reference Architecture—Part 4: Security and Privacy.

[55] ISO/IEC 24392—Security Reference Model for Industrial Internet Platform.

[56] ISO/IEC PWI 27046—Big Data Security and Privacy—Implementation Guidelines.

[57] ISO/IEC 27071—Security Recommendations for Establishing Trusted Connections Between Devices and Services.

[58] ISO/IEC 27090—Artificial Intelligence—Guidance for Addressing Security Threats in Artificial Intelligence Systems.

[59] ISO/IEC 27400—IoT Security and Privacy—Guidelines.

[60] ISO/IEC 27402—IoT Security and Privacy—Device Baseline Requirements.

[61] ISO/IEC 27403—IoT Security and Privacy—Guidelines for IoT Domotics.

[62] ISO/IEC 27404—IoT Security and Privacy—Cybersecurity Labelling Framework for Consumer IoT.

[63] ISO/IEC 29100: 2011, Information Technology—Privacy Framework.

[64] ISO/IEC 29190: 2015, Information Technology—Privacy Capability Assessment Model.

[65] ISO 22301—Security and Resilience—Business Continuity Management Systems—Requirements.

[66] ISO 22313—Security and Resilience—Business Continuity Management Systems—Guidance on the Use of ISO 22301.

[67] ISO 22316—Security and Resilience—Organizational Resilience—Principles and Attributes.

[68] ISO 22318—Security and Resilience—Business Continuity Management Systems—Guidelines for Supply Chain Continuity Management.

[69] ISO 28001—Security Management Systems for the Supply Chain—Best Practices for Implementing Supply Chain Security, Assessments and Plans—Requirements and Guidance.

[70] ISO/IEC 17000 Conformity Assessment—Vocabulary and General Principles.

[71] ISO/IEC 17021-1 Conformity Assessment—Requirements for Bodies Providing Audit and Certification of Management Systems—Part 1: Requirements.

[72] BS 7799 Part 1, Code of Practice for Information Security Management, 1995 and 1998 editions.

[73] BS 7799 Part 2, Specification for Information Security Management System, 1999.

[74] BS 7799 Part 3, Information Security Risk Management, 2006 and 2017 editions.

[75] NIST Special Publications—800 (Guidance) Series, https://csrc.nist.gov/publications/sp800.

[76] NIST Special Publications—1800 (Practice Guides) Series, https://csrc.nist.gov/publications/sp1800.

[77] NIST Cybersecurity White Papers, https://csrc.nist.gov/publications/cswp.

[78] OECD, "The Promotion of a Culture of Security for Information Systems and Networks in OECD Countries," OECD Digital Economy Papers, No. 102, OECD Publishing, 2005.

[79] OECD, "The Role of the 2002 Security Guidelines: Towards Cybersecurity for an Open and Interconnected Economy," OECD Digital Economy Papers, No. 209, OECD Publishing, 2012.

About the Author

Edward Humphreys is an information security management authority and leading-light, and pure mathematician. He has worked in the field of information security, ICT security and risk management for more than 41 years. During this time, he has been an advisor to major international companies and organizations around the globe, and for several European institutions and governments. He is also a visiting professor of ISMS, cyber resilience, risk management and risk psychology, attached to several universities in Europe and Asia.

He has been the Convenor of the ISO/IEC JTC 1/SC 27 responsible for the development and maintenance of the series of ISO/IEC 27001 ISMS standards for more than 34 years. He is often quoted as "the father of ISO/IEC 27001 and ISO/IEC 27002" by ISO and other international organizations, for his original and pioneering work on the original UK versions of these standards as well as his seminal work and leadership in the progression of these UK standards into a set of world best-selling ISO standards. He is also recognized for his work in championing the development and promotion of ISO/IEC 27001 certification around the globe. During this formative time he has been at the forefront in building the international certification framework—from the development of the original accreditation requirements documents for ISMS CBs, through to supporting organizations to become development of CBs, acting as a lead assessor and senior advisor for several National Accreditation bodies, developing the original ISMS competence requirements for auditors and providing professional training on certification for end users, CB auditors and AB assessors. In addition, he is a leading authority on applied information security risk and incident management, and cyber resilience, in particular, related to the challenges of today's digital world.

He was honoured with several prestigious awards, including in 2002 the Secure Computing Lifetime Achievement Award for his achievements on the internationalization of the information security standards based on ISO/IEC 27001. In 2014 he was awarded the prestigious UK Wolfe-Barry gold medal award for his outstanding contributions and services to international standards.

Index

A

Access control
- about, 92 93
- authorization, accountability and ownership, 94
- methods, 94
- models, 93
- registration and deregistration process, 94–96
- rights and information classifications, 93–94
- termination process, 96

Accreditation
- about, 203
- accredited CB selection and, 204
- body, 202
- defined, 202
- standards, 204–5
- *See also* Certification

Action plans, 185

Annex A
- about, 23
- control categories, 104–5
- controls, 103–5
- illustrated, 104
- purpose and definition, 103

Appetite, risk, 42–44

Artificial intelligence (AI), 2, 3

Attitude, risk, 42

Audit report, 208

Audits
- certification scope, 205–6
- competence and, 211–12
- defined, 202
- guidance, standards, 205
- internal, 181–82
- measurement, 167–68
- nonconformities and, 208
- operational, 120
- processes, 206–7
- recertification and, 209–10
- surveillance, 208–9

Audit team, 207

Audit trails, 210–11

Authorization, accountability and ownership, 94

Automated data collection, processing and learning, 175–76

Availability problem, 116–18

Awareness
- communications and, 185–86
- education and, 7
- human resources and, 99–100
- improvements to, 194
- ISMS operations and, 109
- leadership and support and, 67–68
- management policy and, 62–63
- monitoring and review and, 158–59
- policies and procedures and, 82–83
- *See also* Training and awareness

B

Backup(s)
- about, 120
- information identification for, 90
- managed services, 121–22
- methods, 90
- in operational environment, 120–23
- policy, 121
- processes, 122–23
- protection, 91
- records, 92
- restored, testing, 91
- roles and responsibilities, 121
- storage, 91
- technology, 121

Business case, ISMS, 25–26
BYOD (bring your own device), 32, 34

C

Capacity planning, 120, 123–24
Case studies
- example risk levels, 47
- heat map examples, 51
- information classification and access rights, 95
- nonconformity, 192–93
- risk acceptance heat map, 50
- risk scale descriptors, 49
- supply chain requirements, 32
- supply chains, 29

Center for Internet Security (CIS), 75
Certification
- about, 197–98
- accreditation and, 202, 203
- accredited CB selection and, 204
- body, 202
- competence and, 211–12
- costs and resources, 199–201
- defined, 201–2
- global take-up, 198–99
- interested parties involved in, 202
- international, 198–201
- motivation, 199
- recertification and, 209–10
- scope, 205–6
- standards, 205

Change management
- about, 124, 151–52
- change types and, 124–25
- process, 120, 152

Change(s)
- about, 150
- certainty of, 150–51
- forms of, 151
- resistance to, 151–52
- service provider, 163–64
- tracking and reviewing, 153–54
- types of, 124–25, 153

Cloud Control Matrix (CCM), 75
Cloud Security Alliance (CSA), 75
Competence
- certification and audit process and, 211–12
- improvements to, 194
- monitoring and review, 157

Compliance requirements, 35
Conflict resolution, 129
Continual improvement, 187–90
Contract termination, 164
Control examples
- access control, 92–96
- human resource, 99–101
- information-handling, 83–92
- overview, 83
- resource use, 96–99

Controls
- Annex A, 103–5
- determining, 57, 71–75
- effectiveness, 49, 55, 73
- existing, analysis of, 49–50
- framework, 71–72, 76
- incident/attack indications and, 73
- legal, regulatory and contractual, 79
- monitoring and review, 161–62
- operational and business communications, 79
- organizational and management procedures and processes, 77–78
- organizational policy, governance, risk and compliance, 77

personnel, 78
physical, 80
policy and procedure, 80–83
sector and application specific, 101–3
sources of, 74–75
system of, 76–80
technical, 79–80
treatment options and, 73
Control sets
audit trails and, 210
determination process, 72–74
existing, 75

D

Decision-making, informed, 154–55
Deletion and destruction, information, 89–90

E

Effectiveness
holistic, 189–90
information processing, 85
ISMS, 5, 23, 27, 149–50
ISMS controls, 49, 55, 73
maintaining, 189
risk treatment plan, 110
Electronic accounting system, 142–43
Electronic communication and messaging, 87–88
Employment activities, 101, 102
End-user organizations, 205
EU Cybersecurity Act, 8
EU Directive on Electronic Signatures, 8

F

Fax, 88

G

Gap analysis, 173–75
General Data Protection Regulation (GDPR), 75
Government payment system
existing and future deployment, 143–44
introduction to, 143
project scale, 144
See also ISMS use examples

H

Holistic effectiveness, 189–90
Human resources
about, 96–97
acceptable use of, 96–99
general policy provisions, 97–98
leadership and, 65–66
policy procedures and process, 99–100
specific policy provisions, 98–99
training and awareness, 99–100

I

Identify, detect and report, 130
Improvement(s)
about, 215–16
to awareness and training, 194
continual, 187–90
effectiveness, suitability, adequacy and, 189
holistic effectiveness and, 189–90
making, 193–95
to management leadership commitment, 194
nonconformance and, 190–93
planning and implementing, 193–95
to policies and procedures, 195
to processes, 194–95
to resources and competence, 194
Incident management
analysis and evaluation, 131
events, 127
identify, detect and report, 130
incident cause and effects and, 134
investigations, 134
measurement, 171–72
planning, 129–30

Incident management (continued)
post-incident review and follow-up, 133–34
processes, 129–34
recovery, 133
reports, 181
response, 131–32
standards, 136
team, 135
timeliness of response, 132–33
use cases, 128–29
See also ISMS operations
Information
backups, 90–92
classification of, 84
deletion and destruction, information, 89–90
dependency on, *xv*
gathering of, 85
global connection and, 3–4
historical examples of, 1–2
importance of, 1–3
life cycle, 214–15
off-site processing of, 85–86
ownership and accountability of, 84
processing of, 85
protection in operational environment, 119–20
records management, 92
risks to, 2
sharing, exchange, distribution, 87–89
storage, 86–87
use and application, 4
use and misuse of, 84
Information-handling procedures
about, 83
information backups, 90–92
information destruction and deletion, 89–90
information gathering, creation and processing, 85–86
information management, 84
information records management, 92
information sharing, exchange and distribution, 87–89
information storage, 86–87
See also Control examples
Information security
as global issue, *xv*
incidents, 127
managing, 18
policies, 81–83
risk appetite and tolerance, 43–44
Information security management system (ISMS)
about, *xvi*, 5
accredited certification and auditing standards, 14
audits, 205–12
awareness and education and, 7
basis of, 5
business case, 25–26
as business enabler, 216–17
connectivity and outsourcing/cloud and, 31
controls. *see* controls
customers and business partners and, 31
effectiveness and suitability, 149–50
emergence of, 17–18
external issues and context and, 28–30
gathering requirements relevant to, 30–35
interested parties, 30
internal audits, 181–82
internal issues and context and, 27–28
internal technology requirements, 32–34
leadership, 63–64
as living system, 213–14
management and, 7
management policy, 61–63
measurements, 165
needs and expectations, 30–35
organizational context, 26–30
performance and improvements, 214–15

requirements, 13, 214
requirements relevant to, 30–34
resources, 65–66
roles and responsibilities, 64–65
series, standards related to, 15–17
stages, 21
stakeholders and investors and, 30
standards, 12–13
supporting guidelines, 13–14
training, 67–68
Information Society, 3
Informed decision-making, 154–55
Insider threat
about, 114–15
managing, 115–16
reasons for, 114–15
See also Threats, operational
Interested parties, ISMS, 30
Internal audits, 181–82
International certification
costs and resources, 199–201
global take-up, 198–99
motivation, 199
See also Accreditation; Certification
International Electrical
Committee, 11–12
International Organization for
Standardization (ISO), 11–12
International Security Forum (ISF), 75
Internet of Things (IoT), 2, 3
Investigations, incident, 134
ISMS availability and business
continuity
business impact, 137–38
plans, 138–39
processes, 139
standards, 139–40
value and importance of, 136–37
See also ISMS operations
ISMS operations
availability and business
continuity, 136–40
incident management, 127–36
planning, management and
control, 107–9
process examples, 119–27
risk management, 109–10
risk treatment, 110
threat examples, 111
training, awareness and usage, 109
use examples, 140–48
ISMS risk manager, 68–69
ISMS risks
about, 27–28, 44
assessment, 45–52
controls determination, 57
management process, 45–58
monitoring, 59
ongoing reassessment, 58–59
reviews and reassessments, 58–59
risk owner's duties and, 58
risk treatment plan and, 58
statement of applicability, 57–58
treatment, 52–57
treatment, updating, 59
ISMS scope
considerations and inclusions,
35–36
defining, 35, 37–38
example, 38–39
external and internal
connections, 40
ISMS inclusions and, 38
object of, 36–37
scenarios, 36
scope, 34
See also Information security
management system (ISMS)
ISMS use examples
electronic accounting
system, 142–43
government payment
system, 143–44
legal services, 141–42
manufacturing systems, 146–47
outsourcing call center
operations, 144–45
SME design services, 140–41
supply-chain management,
147–48
ISO/IEC 17000 series of standards, 201,
205

ISO/IEC 27000 series
- applications and services, 16–17
- ICT readiness, IDPS, and digital investigations, 16
- ISMS certification and auditing standards, 14
- ISMS controls, 14
- ISMS-related standards, 15
- ISMS requirements, 13
- ISMS supporting guidelines, 13–14
- *See also specific ISO/IEC 27000 standards*

ISO/IEC 27001
- Annex A, 23, 103–5
- certified business environment, 9
- conformance to, 23–24
- emphasis of, 6
- history of, 13, 17–19
- international benchmark, 9
- internationalization, 18
- international standards, 12–13
- ISMS emergence and, 17–18
- in managing information security risks, 5–6
- mandatory statements of requirement, 20–21
- operations, 23
- performance evaluation, 23
- processes, 21
- risk-based approach, 22
- support, 22

ISO/IEC 27001:2022
- about, 13, 19–20
- Annex A, 23, 103–5
- audience, 20
- introduction to, 19–20
- ISMS stages, 21
- mandatory statements of requirement, 20–21
- operations, 23
- organization and top management, 22
- performance evaluation, 23
- processes, 21
- risk-based approach, 22
- support, 22
- *See also* ISO/IEC 27001

ISO/IEC 27002, 14
ISO/IEC 27003, 14
ISO/IEC 27004, 14
ISO/IEC 27005, 14
ISO/IEC 27006, 14
ISO/IEC 27007, 14
ISO/IEC 27008, 14
ISO/IEC 27011, 14, 101
ISO/IEC 27014, 15
ISO/IEC 27016, 15
ISO/IEC 27017, 14, 101
ISO/IEC 27018, 14, 102
ISO/IEC 27019, 14, 102
ISO/IEC 27031, 16, 140
ISO/IEC 27032, 16
ISO/IEC 27033, 16–17
ISO/IEC 27034, 17
ISO/IEC 27035, 16, 136, 140
ISO/IEC 27036, 17
ISO/IEC 27037, 16, 136
ISO/IEC 27038, 136
ISO/IEC 27040, 17
ISO/IEC 27041, 16, 136
ISO/IEC 27042, 16, 136
ISO/IEC 27043, 16, 136
ISO/IEC 27050, 16, 136
ISO/IEC 27799, 102
ISO/IEC JTC 1/SC 27, 12
ISO/IEC standardization, 11–12

L

Leadership, 63–64
Legal, regulatory and contractual controls, 79
Legal and contractual compliance, 164–65
Legal services, 141–42
Legislation and regulation, 7
Life cycle, information, 214–15

M

Malware, 111–13, 120
Management policy
- about, 61

approval, communication, awareness, 62–63
elements, 61–62
policy review, 63
See also Policies and procedures

Management reviews
about, 182–83
action plan, 185
input for, 183–84
output of, 184–85
recommendations for improvement, 184–85

Manufacturing systems, 146–47

Measurement programme
automated data collection, processing and learning, 175–76
data gathering and collection, 170–71
gap analysis and benchmarking methods, 173–75
incident management, 171–72
measurement methods, 169–70
measurements in practice, 176–77
measurement templates, 171
objectives, 168–69
protection of data, 170
scorecards and questionnaires, 172
tests and on-the-job exercises, 173

Measurements
about, 165–66
examples of areas of, 166–68
human resources, 167
management and user controls, 167
management processes, 166
methods, 169–70
in practice, 176–77
reviews and audits, 167–68
risk(s), 179–80
security processes, 167

Metrics, 165, 179–80

Monitoring and review
about, 155
awareness, training and career development, 158–59
business demands and requirements, 156
competence, 157
information security controls, 161–62
information security processes, 159–65
information security roles, 159
legal and contractual compliance, 164–65
performance, 155
security culture, 157–58
staff, 158
technology and network services, 162
third-party contracts and services, 163–64

N

National Cyber Security Centre (NCSC), 75

NIST
availability and business continuity standards, 140
Computer Security Resource Center, 74
incident management standards, 136
risk management standards, 68

Nonconformance
about, 190
audits and, 208
case studies, 192–93
common causes, 191
corrections, 190
corrective actions/root causes and, 190–91
See also Improvement(s)

O

On-site/off-site storage, 87
Operational audits, 120
Opportunity, 42
Organization, this book, *xvi*

Organizational context
business understanding and, 26–27
external issues and, 28–30
internal issues and, 27–28
See also Information security management system (ISMS)
Outsourcing call center operations, 144

P

Paper-based storage, 86–87
Performance evaluation
awareness and communications, 185–86
change and certainty of change, 150–51
change management, 151–55
internal audits, 181–82
management reviews, 182–85
measurements programme, 165–77
monitoring and operational reviews, 155–65
ongoing risk management, 177–81
Personnel controls, 78
Physical controls, 80
Policies and procedures
about, 80
approval, communication, awareness, 82–83
improvements to, 195
purpose of, 81–82
review, 83
See also Controls
Policy review, 63
Post and courier services, 88–89
Processes
availability and business continuity, 139
backups, 120–23
capacity planning, 120, 123–24
change management, 120, 124, 152
example, 108
examples of sketches of, 119–27
improvements to, 194–95
incident management, 129–34
information protection, 119–20
information security, monitoring and review of, 159–65
logistics and supply chain management, 125
planning, management and control, 107–8
reviewing, 160
risk management, 45–58
third-party services, 125–27
tracking and reviewing change, 153–54

Q

Questionnaires, 172

R

Recertification, 209–10
Records management, 92
Recovery, 133
Recruitment activities, 100
Registration and deregistration process, 94–96
Requirements
compliance, 35
in contracts and SLAs, 35
gathering, ISMS and, 34–35
internal business, 34
internal technology, 32–34
ISMS and, 30–34
supply chain, 32
Resources
about, 96–97
acceptable use of, 96–99
certification and, 199–201
general policy provisions, 97–98
improvements to, 194
leadership and, 65–66
policy procedures and process, 99–100
specific policy provisions, 98–99
See also Human resources
Risk acceptance heat map, 50–52
Risk-adverse, 42
Risk assessment

acceptance criteria, 45–46
analysis and, 46–48
analysis of existing ISMS controls and, 49–50
approach to, 4–5
evaluation and, 50
evaluation decision-making and, 50–52
frequency spectrum, 179
identification and, 46
ongoing, 109–10
process, 45
regular (planned), 178–79
risk table scales and, 48–49
See also Risk management process

Risk avoidance, 53, 54

Risk levels, 47

Risk management
about, 215
measurements and metrics, 179–80
ongoing, 177–81
regular risk assessments, 178–79
reviews, 180–81
risk responsiveness and commitment, 177–78

Risk management process
assessment, 45–52
changes in, 45
controls determination, 57
owner's duties, 58
reviewing, 160
statement of applicability, 57–58
treatment, 52–57
treatment plan, 58
See also Processes

Risk monitoring, 160–61

Risk responsiveness/commitment, 177–78

Risk(s)
about, 41–43
analysis, 46–48
analysis of existing ISMS controls, 49–50
appetite, 42–44
attitude, 42
controls determination, 57
defined, 41–42
disruption from, 56
evaluation, 50
evaluation decision-making, 50–52
external, 28–30
identification, 46
information, 2
information security, 5–6, 43–44
ISMS, 27–28, 44–59
managing. *see* risk management
monitoring, 59
opportunity and, 42
reducing consequences of, 56–57
reducing likelihood of, 55–56
retaining/accepting, 55–57
reviews and reassessments, 58–59
tolerance, 42, 43–44
transferring, 54–55

Risk table scales, 48–49

Risk treatment
about, 52–53
cycle, 53
ongoing, 110
options, 53–54
reducing consequences of risks and, 56–57
reducing likelihood of risks and, 55–56
retaining/accepting risks and, 57
risk avoidance and, 54
transferring risks and, 54
updating, 59

Risk treatment plan, 58

Roles and responsibilities, 64–65

S

Sarbanes Oxley (SoX or Sarbox), 9

Scorecards, 154, 172

Sector and application specific controls, 101–3

Service-level agreements (SLAs), 126–27

SMS design services, 140–41

Social engineering, 118–19

Standards, 136

Statement of Applicability (SoA), 58
Storage, information, 86–87
Supply-chain management
  assessment, 128
  existing and future ISMS deployment, 148
  introduction to, 147
  process, 120
  third-party logistics and, 125
  use-case example, 147–48
Supply chains, case studies, 29, 32
Surveillance audits, 208–9
System availability, 116–18
System of controls
  control framework, 76
  examples, 77–80
  requirements, 76–77
  See also Controls

T

Technical controls, 79–80
Technology and network services, monitoring and review, 162
Termination process, user access, 96
Third-party services
  monitoring and review, 163–65
  service agreements, 126–27
  supplier relationships and service management, 163
  supply-chain management and, 125
  uses, 125–26
Threats, operational
  examples, 111–19
  insider threat, 114–16
  malware, 111
  social engineering, 118–19
  system availability, 116–18
  unauthorized access, 113–14
Tolerance, risk, 42, 43–44
Training and awareness
  human resources and, 99–100
  improvements to, 194
  ISMS operations and, 109
  leadership and support and, 67–68
  monitoring and review and, 158–59
Trojan horse, 112, 113

U

UK legislation, 8
Unauthorized access, 113–14
US cyber security-related regulations, 8

V

Viruses, 111–13
Voice communication, 88
Vulnerability management, 120

Artech House
Information Security and Privacy Library

Rolf Oppliger, Series Editor

Attribute-Based Access Control, Vincent C. Hu, David F. Ferraiolo, Ramaswamy Chandramouli, and D. Richard Kuhn

Biometrics in Identity Management: Concepts to Applications, Shimon K. Modi

Bitcoin and Blockchain Security, Ghassan Karame and Elli Androulaki

Blockchain for Cyberphysical Systems, Ali Dorri, Salil Kanhere, and Raja Jurdak

Bluetooth Security, Christian Gehrmann, Joakim Persson, and Ben Smeets

Computer Forensics and Privacy, Michael A. Caloyannides

Computer and Intrusion Forensics, George Mohay, et al.

Contemporary Cryptography, Second Edition, Rolf Oppliger

Cryptography for Security and Privacy in Cloud Computing, Stefan Rass and Daniel Slamanig

Defense and Detection Strategies against Internet Worms, Jose Nazario

Demystifying the IPsec Puzzle, Sheila Frankel

Developing Secure Distributed Systems with CORBA, Ulrich Lang and Rudolf Schreiner

Electric Payment Systems for E-Commerce, Second Edition, Donal O'Mahony, Michael Peirce, and Hitesh Tewari

End-to-End Encrypted Messaging, Rolf Oppliger

Engineering Safe and Secure Software Systems, C. Warren Axelrod

Evaluating Agile Software Development: Methods for Your Organization, Alan S. Koch

Fuzzing for Software Security Testing and Quality Assurance, Second Edition, Ari Takanen, Jared DeMott, Charlie Miller, and Atte Kettunen

Homeland Security Threats, Countermeasures, and Privacy Issues, Giorgio Franceschetti and Marina Grossi, editors

Identity Management: Concepts, Technologies, and Systems, Elisa Bertino and Kenji Takahashi

Implementing Electronic Card Payment Systems, Cristian Radu

Implementing the ISO/IEC 27001 ISMS Standard, Third Edition, Edward Humphreys

Implementing Security for ATM Networks, Thomas Tarman and Edward Witzke

Information Hiding Techniques for Steganography and Digital Watermarking, Stefan Katzenbeisser and Fabien A. P. Petitcolas, editors

Internet and Intranet Security, Second Edition, Rolf Oppliger

Introduction to Identity-Based Encryption, Luther Martin

Java Card for E-Payment Applications, Vesna Hassler, Martin Manninger, Mikail Gordeev, and Christoph Müller

Medical Device Cybersecurity for Engineers and Manufacturers, Axel Wirth, Christopher Gates, and Jason Smith

Multicast and Group Security, Thomas Hardjono and Lakshminath R. Dondeti

Non-repudiation in Electronic Commerce, Jianying Zhou

Outsourcing Information Security, C. Warren Axelrod

The Penetration Tester's Guide to Web Applications, Serge Borso

A Practical Guide to Managing Information Security, Steve Purser

Privacy Protection and Computer Forensics, Second Edition, Michael A. Caloyannides

Role-Based Access Control, Second Edition, David F. Ferraiolo, D. Richard Kuhn, and Ramaswamy Chandramouli

Role Engineering for Enterprise Security Management, Edward J. Coyne and John M. Davis

Secure Messaging on the Internet, Rolf Oppliger

Secure Messaging with PGP and S/MIME, Rolf Oppliger

Securing Information and Communications Systems: Principles, Technologies, and Applications, Steven Furnell, Sokratis Katsitas, Javier Lopez, and Ahmed Patel

Security for Cloud Computing, Roberto Di Pietro and Flavio Lombardi

Security Fundamentals for E-Commerce, Vesna Hassler

Security in Wireless LANs and MANs, Lakshminath R. Dondeti and Thomas Hardjono

Security Technologies for the World Wide Web, Second Edition, Rolf Oppliger

Signal and Messaging Layer Security, Rolf Oppliger

SSL and TLS: Theory and Practice, Second Edition, Rolf Oppliger

Techniques and Applications of Digital Watermarking and Content Protection, Michael Arnold, Martin Schmucker, and Stephen D. Wolthusen

Understanding Voice over IP Security, Alan B. Johnston and David Piscitello

User's Guide to Cryptography and Standards, Alexander W. Dent and Chris J. Mitchell

For further information on these and other Artech House titles, including previously considered out-of-print books now available through our In-Print-Forever® (IPF®) program, contact:

Artech House	Artech House
685 Canton Street	16 Sussex Street
Norwood, MA 02062	London SW1V 4RW UK
Phone: 781-769-9750	Phone: +44 (0)20 7596-8750
Fax: 781-769-6334	Fax: +44 (0)20 7630-0166
e-mail: artech@artechhouse.com	e-mail: artech-uk@artechhouse.com

Find us on the World Wide Web at: www.artechhouse.com